T0288995

Shifting
Safety and Health
Paradigms

F. David Pierce

Government Institutes
Rockville, Maryland

Government Institutes, Inc., 4 Research Place, Suite 200, Rockville, Maryland 20850

Copyright ©1996 by Government Institutes. All rights reserved.

00 99 98 97 96   5   4   3   2   1

No part of this work may be reproduced or transmitted in any form or by any means, electronic or mechanical, including photocopying, recording, or any information storage and retrieval system, without permission in writing from the publisher. All requests for permission to reproduce material from this work should be directed to Government Institutes, Inc., 4 Research Place, Suite 200, Rockville, Maryland 20850.

The reader should not rely on this publication to address specific questions that apply to a particular set of facts. The author and publisher make no representation or warranty, express or implied, as to the completeness, correctness, or utility of the information in this publication. In addition, the author and publisher assume no liability of any kind whatsoever resulting from the use of or reliance upon the contents of this book.

Library of Congress Cataloging-in-Publication Data

Pierce, F. David

     Shifting safety and health paradigms / F. David Pierce.
         p.   cm.
     Includes index.
     ISBN: 0-86587-527-8
     1. Industrial safety--United States.   I. Title
  T55.P53   1996
  363.11' 0973--dc20
                           96-19082
                              CIP

Printed in the United States of America

TABLE OF CONTENTS

ABOUT THE AUTHOR

F. David Pierce is a Certified Safety Professional and Certified Industrial Hygienist with over 25 years experience in industry and government. He is the manager of safety, environment, and security at Westinghouse Electric Corporation's Western Zirconium Plant, where he spearheaded the efforts which resulted in a Malcolm Baldridge National Quality Award. Pierce is also director of Strategic Solutions, Inc., a safety and environmental management firm.

Pierce is the author of *Total Quality for Safety and Health Professionals* (Government Institutes, 1995) and *New Directions/Dynamic Strategies for Safety and Health*, a correspondence course sponsored by the Professional Alliance to Business. He is a frequent speaker at professional conferences and teaches popular courses on Total Quality and Occupational Safety and Health Management for the National Institute for Occupational Safety and Health's Education Resource Centers.

Pierce received his Master of Science degree in public health from the University of Utah–Salt Lake City. He is past president of the Utah chapter of the American Society of Safety Engineers and former chair of the Utah Occupational Safety and Health Advisory Council. Pierce currently chairs the Utah Safety and Hygiene Conference Committee and has served on the advisory committee for the Rocky Mountain Center for Occupational Safety and Health. He is a member of the American Industrial Hygiene Association, American Society of Safety Engineers, American Academy of Industrial Hygiene, International Hazardous Materials Association, and National Association of Environmental Professionals.

DEDICATION

To my family, who puts up with my ranting and ravings and still finds ways to be proud of me, and especially my wife, Gayle who supports my free thinking and continually encourages me to chase my dreams. To those fellow professionals who can, are beginning to, or have already changed their paradigms to think differently and see how to be successful in their future. To my friends and allies, Tim Razzeca, Carlos Aguilar, Bob Cunliffe and Ed Kiely who not only share tomorrow's paradigms with me, but know it is the future. To facility managers like Dick Gerwels and Bill Whitehead who have the courage to practice the new management paradigm although, too often, it is outside of current tradition. Also to my many "teachers" throughout my life who, through their counsel and practice, taught me to reach beyond tradition and accepted thought. And to Mrs. Smith, my sixth grade teacher—I told you so.

PREFACE

21 EXAMPLES OF HOW TODAY'S SAFETY AND HEALTH SYSTEM ISN'T WORKING

sys•tem (sis'təm) n.[LL. *systema* <Gr. *systēma* (gen. *systēmatos*) <*synsistanai*, to place together < *syn-*, together + *histanai*, to set: see STAND] 1. a set or arrangement of things so related or connected as to form a unity or organic whole *[a solar *system*, school *system*, *system* of highways]* [1]

sys•tem for work•er safe•ty and health (sis'təm fôr wʉr'kər sāf'tē ənd helth) n. 1. a set of parts so related or connected as to form a unity for providing a working environment that is free from danger, injury and maintains physical and mental well-being, freedom from disease for those persons employed to do physical or mental work for wages

---

[1] Webster's New World Dictionary of the American Language, Simon and Schuster, New York, NY, 1982.

#1

A new employee arrived for his first day of work. Wearing his newly purchased steel-toed boots, the nineteen-year old was excited. He had just taken a quantum leap in earning power, moving from fast food to construction. He knew that he would be starting at the bottom, but the prospect of working his way up the ladder was a career decision he was excited about. Unfortunately, his career only lasted three hours. Working at his first assignment, cleaning out the bed of a large earthmover with a water hose, he had crawled between the scoop bed and the propped open rear unloading gate. While his attention was directed at the water-hose stream cleaning dirt from the bed, for some reason the block that was holding up the unloading gate failed. The falling gate caught him between the gate and the bed and cut him in half.

#2

An electrical apprentice had been working on a job for two days. One task required that he descend into an underground conveyor alley to wire a new drive box. A conduit was needed to contain the lengthy run of wire back to the control station. He had been busy installing the conduit most of the morning, moving slowly down the conveyor alley toward the drive box location. One of the known challenges in this conveyor alley was the high ground-water level. It was dirty water with a very high concentration of suspended solids. Stepping into a deep puddle at one point in the alley, the new apprentice took out his electric drill to secure the conduit. With one contraction of the trigger, the current from his non-insulated tool found a new path for the 110-volt current—through his body. It stopped his heart. After a three-hour search when he didn't show up for his ride home, his body was found. He was nineteen and engaged to be married. He had been employed by his father's electrical contracting company. His father was the one who found him.

#3

A certain worker had a reputation for complaining a lot about safety conditions and lack of standardized safety rules for performing hazardous jobs. He had previously worked for a very safety-conscious company. Unfortunately, because of market competition, that company had gone through some rough times and the worker had been laid off. He was very thankful to find this new job, but the safety conditions were terrible compared to his previous job. His pro-safety attitude was not appreciated by his new employer or his fellow employees. Over the nine months he worked there, he gained the reputation of being a whiner. His regular complaints about unsafe working conditions were constantly slowing production down. His supervisor talked with him many times about his "bad attitude." Finally, after refusing to go into a confined space alone without any testing or precautions, he was fired for insubordination. He fought the discharge through normal labor relations channels only to meet mountains of documentation about being counseled for his negative attitude. His discharge was upheld.

#4

Three bricklayers at a large steel-making facility were rebricking a basic oxygen furnace. The furnace, a large vessel approximately thirty-feet-high and ten-feet-around inside, was formed like a large brandy snifter, open at the top. The bricklayers gained access to the job by climbing down into the vessel on a ladder; brick was placed in the vessel by crane.

The bricklayers had been working in the vessel and climbing out for breaks for most of the day. As with most complex jobs in the steel mill, there were detailed procedures that outlined the steps to be taken before, during, and after the job. The workers knew that the vessel was a confined space. They knew that. As a precautionary

measure to ensure that there was adequate oxygen in the space, the
hard-piped oxygen line to the bottom of the furnace was left in place;
like most valves in such a hostile environment, it leaked. Because
oxygen is heavier than the nitrogen that makes up more than 70
percent of the air we breathe normally, it had pushed the air up and
out of the vessel. The atmosphere in the vessel was well enriched
with oxygen. As the day was getting late, the bricklayers were tired
of always climbing out of the vessel for their breaks. One decided to
take his break inside the vessel. When he struck a match to light his
cigarette in the oxygen-rich atmosphere, all three men became fuel
for a rapid and hot fire. The combined losses from the fire were
devastating to three wives, seven children and nine grandchildren.

#5

An OSHA inspector visited a medium-sized metal product
manufacturing facility. After conducting a day-long, wall-to-wall
scheduled compliance inspection, the inspector noted several items
that the employer would probably receive citations for. These
included improperly labeled chemical-product containers,
inadequately labeled electrical disconnections, and a hole in a
catwalk surface. The inspector also had questions about some injuries
that weren't recorded correctly on the OSHA 200 Log. Twenty-three
days later, an employee was killed at this same facility when he failed
to lockout a large piece of metal forming equipment, and was pulled
into the metal forming rolls. An OSHA investigation of the death
found that the facility had no lock out program or practice. It also
found that the facility had several confined spaces entered by
employees routinely, although there was no confined space entry
program or permit system. Unfortunately, the first OSHA inspector
hadn't asked about lockout or confined spaces during the scheduled
inspection; these items were absent from his inspection notes. The
family of the dead employee surely wished lockout had been covered
in that first inspection.

#6

    A large insurance company that writes a considerable amount of workers' compensation coverage for employers recently expanded its staff. It added eight investigators to look into cases where workers' compensation fraud was suspected. This carrier believed that as much as 35 percent of its direct costs related to workers' compensation benefits were the result of fraudulent claims. Of these fraudulent cases, more than 75 percent were back injury cases that had lasted well over two years. Each of these cases easily cost the insurance company an amount equal to the salary and benefits of one investigator. Thus, to the insurance company, investigating such cases was a cost-containment strategy that made good economic sense.

#7

    An employee who operated a switching engine at a large metal manufacturing facility did so by remote control. He was killed as he was trying to hook up one car to another at a loading dock. While walking ahead of the oncoming engine and short string of cars, he was caught by the corner of the first oncoming car and rolled between the dock and the car. Obviously, the string of cars had arrived sooner than he expected. The four inches of clearance between the car's side and the loading dock gave him little room to escape. No OSHA citation was issued; the worker had just been careless.

#8

    Another employee of this same facility was working as a switchman on the transportation crew that moved materials around the plant via a railroad. On this day, he was part of a crew of two, himself and an in-cab engineer. They had been collecting up stray cars that were scattered around the facility so that they could be returned to the railroad yard for staging. They had already hooked up about 11 cars and had another three or four to get before they completed their afternoon's work. Because of the length of the string of cars, it was customary to communicate by radio. With the switchman at one end of the train and the engineer at the other, the switchman could direct the engineer's movement of the string of cars. This day was no different. Walking at the end of the slowly moving string of cars, the switchman directed the train movement toward another stray car. For coupling to occur between two cars, the couplers of both cars have to be in the open position; when the two couplers meet, both couplers close and attach one car to the next. As the string of cars progressed toward the stray car, the switchman noticed that the coupler on the stray car was in the closed position. Moving quickly, he jogged ahead and attempted to open the coupler. It didn't open with the first or the second try. Kicking the locking pin with his foot, he muscled the coupler open just as the string of cars arrived behind him. The oncoming coupler pushed him into the waiting coupler on the stray car; the two couplers met and closed him inside the coupling jaws. It took more than two hours to disconnect the cars and free the trapped switchman. During this long and delicate effort, he received constant medical care. Miraculously, he survived and lived another fifteen years before he died of problems associated with the injury. This case alone cost the self-insured company more than $2,500,000.

#9

As college students regularly do, a college junior took a summer job with a tree grooming service to earn money for school. Daily, a work crew of two would trim trees, grind the trimmings to pulp, and then dispose of the pulp at the local landfill. They trimmed trees for all sorts of reasons including tree disease, encroachment on neighboring property or against houses, unattractive appearance, etc. One day they went to trim some trees that were growing around some electrical lines that went to a house. It had been raining off and on most of the day. Taking his pole-mounted cropping saw, the young worker reached up into the tree to cut away some of the branches that blocked his view of the lines. Unfortunately, one of the "branches" he began to saw was the electrical line; the wet pole served as an excellent path for the electrical current to go to ground, through his body. Although they worked on him for over a half hour, the attending paramedics were unable to restart his heart. The student had carried a 3.92 GPA in engineering at the university he attended and was the only son of a widowed mother.

#10

Management at this particular company was very autocratic. Workers were there to work and not get injured. If they were injured, obviously they were careless and were usually disciplined. Everyone knew the game: if you wanted to stay employed, you didn't get injured. One maintenance mechanic slipped while trying to force a piece into a machined part. With great force, his hand impacted on a sharp edge of a mounting bracket, resulting in a deep laceration to the palm. The next week employee evaluations were scheduled. The employee knew very well what effect his injury would have on his imminent evaluation, especially if he were disciplined for carelessness. Wrapping a rag around his injured hand, he told his supervisor that he had to go home and take his wife to the doctor because of an unanticipated emergency. Grumbling, the supervisor let him go but insisted that the mechanic make up the time the next day. The mechanic went to the emergency room and had the laceration taken care of at his own expense. Neither the company nor the supervisor ever knew that the injury occurred.

#11

The safety policy at a large manufacturing company stated that workers were required to wear safety belts and be tied-off with lanyards whenever they worked unprotected more than six feet above the floor. The policy noted no exceptions to this requirement. In practice, however, several tasks involving elevated work were performed without tie-off because no tie-off means were available or convenient. Due to the company's downsizing efforts, several specialized functions had been eliminated or contracted out to other businesses; one of them was servicing the ceiling-mounted unit heaters. In order to service these heaters, a technician had to work on a ladder 15 to 25 feet above the floor. However, this was one of the jobs where tie-off was not practiced because there was no convenient to do it. Once during the servicing of the company's unit heaters, the ladder that the technician was standing on slipped, causing the technician to fall about 22 feet onto the concrete floor below. His injuries included a concussion, a broken wrist and arm, a broken pelvis, two broken ribs, some internal injuries and a broken back in two places. He spent three months in the hospital and the rest of his life in a wheelchair as a paraplegic. He sued the company for not enforcing the policy. He also sued the company's general manager and the manager of the department where the injury occurred. The suit was settled out of court for more than $1,000,000.

#12

A contractor was welding at a refinery on a vessel that normally contained very volatile explosive liquids at elevated temperatures. The unit had been taken out of service and all lines into the vessel were blanked or isolated using double-block-and-bleed techniques. The job had already taken two days. The company had hoped that only one day would be required, but because there was more corrosion and oxidation of the walls of the vessel than expected, the repair had taken additional time and material. As a result, production was behind, but the company was given assurances that the unit would be back on line sometime on the third day. In preparation for this day's work, all isolation points were checked, testing for explosive gas mixtures was performed, and a hot work permit was completed by the facility safety engineer. Only a small amount of work needed to be done when the contractor crew returned from lunch. Figuring that only "cosmetic" work was left, without telling the contractor crew and against the insistence of the facility safety engineer, the production supervisor ordered his crew to remove the vessel isolation devices, fill the vessel with a volatile liquid mixture and start bringing up the temperature of the vessel. When the contractor crew returned from lunch, one welder lit his torch and touched the wall of the vessel with a cutting flame. The explosion was heard more than two miles away but not by the welder. OSHA citations were issued and suits followed. However, the production supervisor was not disciplined for the incident and one year later he was still on the job.

#13

One day a 32-year-old bricklayer injured his back while lifting an 80-pound bag of cement. As he was shifting the bag up onto his shoulder, he heard something in his lower back "pop." Suddenly he knew he was on the ground with severe pains shooting down his right leg. He had to be taken on a stretcher to the local emergency room where he was diagnosed with a herniated disc in his lower back. He was told to go home and rest in bed. By the following week his back was a little better; he could stand and move around a little. But, any real movement or effort was very painful. The doctor prescribed regular physical therapy. Two weeks later when things hadn't improved measurably, the doctor recommended surgery to fuse the bones of his lower back. The surgery was considered successful, but after three months of therapy and healing, the injured worker seemed to reach a plateau. A year later, the worker was still not able to return to his job; he was still not able to get around xxii / Shifting Safety and Health Paradigms

without considerable pain and couldn't even lift his four-year-old daughter. Maybe he never would be able to work again. But soon after that, a private investigator working for the compensation carrier videotaped the injured man playing golf, carrying bundles of shingles up a ladder and shingling his own roof, and playing football with his son.

#14

A plant was about to achieve a safety record. If they went two more months without a lost-time injury, they would reach 2,000,000 work hours and be number one in the region. The plant manager wanted to be number one very badly; it was important to everyone at the plant. Unfortunately, a young female machine operator was struck by a large material transporting cart that weighed about 2,500 pounds. Her lower leg was broken, requiring casting and immobilization for proper healing. Because of the safety record, pressure was applied so that she would not lose any work days. Doctors were talked to and concessions were made. Each morning, she was picked up at her house by an ambulance and driven to the plant clinic where she answered the phone. Each night the ambulance took her home. This continued until she was able to get around with the help of friends. Yes, the plant made the safety record—at least officially.

#15

A shop where heavy pieces of equipment were made used several overhead cranes and hoists for moving and manipulating the equipment during production. There was no requirement for wearing hard hats in the facility. "The employees don't like hard hats. They think that they are uncomfortable and pose a greater risk than not wearing them," a safety consultant was told during a contracted compliance audit. The company reinforced that belief among workers by regularly giving out baseball hats as employee and team awards for production and quality accomplishments. In fact, almost everyone on the shop floor wore a baseball hat. The plant had been inspected twice in the last five years by OSHA; however, no citation had ever been issued for not requiring hard hats nor was a policy change recommended, although it was well known that moving heavy equipment by crane and hoist posed significant overhead hazards. One day, as a crane was moving overhead, a three-quarter-inch diameter, six-inch-long structural bolt from the crane bridge dropped fifty-three feet. The bolt struck a senior machine operator squarely on the top of the head, killing him instantly. His baseball hat provided him no protection. The dead worker would have retired in four and one half months. At the time of his death, his retirement paperwork was on the plant manager's desk awaiting approval.

#16

A company that was very concerned about regulatory compliance hired a consulting firm to perform a detailed compliance audit and tell them what they needed to fix. Several serious injuries had occurred at the company in the past three months, but they had not been cited by OSHA in more than two years. A team of consultants poured over the plant and found that there were no written programs guiding any safety effort, and that safety was clearly not seen as a management responsibility in the line structure. In their report, the consultants recommended many written programs to guide efforts toward compliance. One was a policy establishing management responsibility for employee safety and deficiency. Another recommendation focused on the importance of management training as an integral part of establishing responsibility for safety culture at the company. A third recommendation concerned building a program of regular team safety inspections of the operations. In a bold move, the company's management hired the consulting firm to write several key compliance programs and the policy to establish management responsibility. As the policy and program drafts began to arrive, the company became concerned about the format that would be used in the documents. A technical writer was hired and many program formats were looked at. However, six months later, no format had been agreed upon, and no written policies establishing management responsibilities and compliance programs had been finalized or approved for issuance. No management training had been provided either. No program of regular team safety inspections had been started. Six months after the initial consultant's report on compliance and recommendations had been delivered to the company, no progress had been made toward achieving the company's stated goal—compliance. The consultants' contract was canceled because it was just too expensive.

#17

A welder was finishing a large job using a portable grinder mounted with a ten-inch abrasive disk. The grinder was air-driven, not electrically-driven. As designed by the manufacturer and as required by company policy, the grinder came with a guard to provide protection for the operator from damaged abrasive stones and disks which can fly off at tremendously high velocities. The guard of the grinder the welder was using, however, had been removed because it was just too inconvenient. Besides, the welders were careful; they always inspected the stones and disks before use. All abrasive stones and disks were stored indoors, away from any moisture and in such a way as to prevent any possible damage. Additionally, as a new stone or disk was mounted on a grinder, it was first brought up to speed inside a "test box" that would contain all debris if the stone or disk shattered. But on this particular day, as the welder was about to finish the job, the disc suddenly shattered. Without warning, large pieces of the fractured disk flew out in all directions. One large piece hit the welder in the right lower abdominal area, just above the leg. Penetrating deeply, it severed major blood vessels. His fellow workers were unable to stop the bleeding or summon help in time. Within a few minutes, the welder was dead from massive blood loss. The investigation revealed that a critical orifice inside the grinder that regulated rpms was clogged due to rust and moisture from the air line that fed it. At the time the disk shattered, the operating speed of the grinder, normally 6,000 RPM, had almost doubled to 11,500 RPM. The disk was only approved for speeds up to 7,500 RPM. The excessive speed caused the disk failure and also increased the velocity and hazard of the flying debris. OSHA cited the company for not having a guard on the portable grinder, a serious finding. The company did not contest the citation and paid the $450 penalty.

#18

A medium-sized manufacturing facility was targeted by the local OSHA office for a wall-to-wall inspection due to their high injury rate. The injury report filed with the OSHA office for the previous year showed a rate for recordable injuries of more than 50 percent. Oddly enough, the lost time injury rate for this same facility that year was a low 1.5 percent. But, because of the high recordable injury rate, the facility was moved up on the OSHA office's compliance inspection schedule. The first action the compliance inspector took when he began his inspection of the facility was to review the current injury and illness log. Again it was high, above 42 percent. The recorded injuries seemed to show no pattern or specific location in the facility. They were happening all over the plant and were injuries to all body parts including backs, fingers, arms, hands, legs, etc. The wall-to-wall inspection revealed a very clean and orderly facility, not what the inspector had expected to find. Only three not serious items were identified during the two-day inspection. But, the company was concerned and called in a safety consultant to take a look. Seeing the number of entries noted on the injury and illness log, the safety consultant quickly deduced what was wrong. The Human Resources representative whose job it was to keep the log didn't know what constituted recordability. She thought that if the injury was seen at

xxvi / Shifting Safety and Health Paradigms

the local emergency clinic or at the hospital, it was recordable. The plant had no medical facilities or trained medical providers on-site. However, the company was very employee-concerned, so if an injured employee wanted a medical person to look at the injury, no matter how minor, they quickly sent him or her to the clinic or hospital. After a review of the medical examination reports, numerous injuries were "lined-out" on the injury and illness log. As a result, the recordable injury rate for the year was reduced to 8 percent, well under the industry sector's average injury rates and in line with the lost time injury rates.

#19

The president of a national organization of safety and health professionals was speaking at a local meeting luncheon. "We're going global," the organization's president stated to the gathering of members. "Excuse me," a low volume request came from one of the members, "could you explain to me what going global is going to do to help me try to improve the safety of the workers at my facility?" "Only through global interaction can the profession gain the respect it deserves and impact global standards for worker safety and health," the president explained. "Yea, but most of the time I'm just fighting change, management, union priorities and not knowing everything about what I'm doing," the member began. "How are going global, respect and global standards going to help me? I'm beginning to feel all alone in the trenches." "You have to broaden the way you think of your job," was the answer from the organization's president. "We can no longer just think "my job" and "my site" if we are going to be a force in tomorrow's global workplace." The member quietly shook his head. He hadn't been heard by the president although everyone else in the room had listened and understood very well.

#20

It was crucial that a chemical reactor's contents be kept in a temperature range just below its boiling point during the manufacture of a shock-sensitive material. The reactor held approximately a ton of the mixture, and production of a batch took about eight hours to complete. The operators that monitored and controlled the process had to check on the temperature of the reactor's contents continually. In order to know if it was too hot or if the rate of reaction was too high, the operators had to refer to a graph that was taped to the wall. Then, based on the elapsed time of reaction, the change in temperature, and the amount of material in the reactor, they had to calculate the probable maximum batch temperature at two or three times during the reaction process. This was important because the process was exothermic and if it was not carefully controlled, the reaction temperatures could get out of control. If the reactants were allowed to boil, the material could detonate. If the temperature was too low, the resulting mixture would not pass the mandatory quality tests. Normal practice was for a batch to be started by one crew and then turned over to the next. On this particular day, the operator going off-shift was in a hurry to leave, misread a temperature at one point in the reaction, and recorded the wrong temperature on the production log. The incoming operator, at a specified point in the reaction, took a correct temperature reading, subtracted the incorrect previous temperature, and proceeded to calculate the probable maximum batch temperature using the wall-mounted graph. The result was too low. As a precaution, he had another operator check his calculations. The calculations were correct, so, according to procedure, the operator raised the heating power to the vessel. Shortly, the process rate took off and the temperature climb could not be controlled. The contents of the reactor came to a boil, and the ton of product detonated. Two people were killed and half of the production facility was leveled.

#21

A facility's injury and illness rate was unacceptably high. Compared to the other plants in the large corporation, this facility's rate was five times higher. Something had to be done. In an effort to turn things around, the present safety specialist was "given another opportunity" and a new safety person was hired. The new person knew very well that upper management expected the rates to go down and that his job security was dependent on that happening. He busily began to write and implement new programs. But, with all the "fires" he had to attend to daily, it was difficult to find time. He spent evenings and at least one day each weekend at the plant working on safety and health programs. But the new programs brought an increased demand for the safety person's time to administer and coordinate them. "I've got to have some help," the safety specialist complained. So the previous safety specialist was brought back to help. The first year saw some improvement in the safety performance. The rate dropped 10 percent. The second year, however, the rate stayed the same and the third year it went up 5 percent. The safety specialist was called into the plant managers office. "Why aren't the rates going down?" he was asked. "I've given you adequate resources and an additional staff person." The safety specialist answered, "I've got to have more help or management has to start taking some of these responsibilities." That wasn't the answer the plant manager wanted, so he let that safety specialist go and began a search for someone who could get the job done.

Each of these examples documents a breakdown in some part of the system for worker safety and health. I have only included twenty-one when, in reality, there are countless examples. I have thousands recorded in my files alone. Is each example significant? Yes and no. Yes, because each represents a problem in the system for worker safety and health that we can learn from, and fix. Like the individual trees in a forest, each problem is significant and unique. But each example is also not significant because after you read a number of them, you begin to be desensitized.

Just as it's important to see each tree, it's also important for us to get a vision of the forest and not take count of all the trees that comprise it.

Too often we look too closely at each problem. We study each example, each failure, each accident, each injury or illness, and devise corrective actions to prevent them from happening again. Theoretically, preventing one problem at a time, over a long period of time, makes all the little problems go away. This approach, however, creates so much "noise" and produces so many "things" to do that most often we contradict earlier strategies or create a subsystem that takes on a life of its own.

This book begins with examples because this is the way we usually see things best. "Give me an example of what you are saying," is a common request from a listener. So, I've started with examples. In this book, each major component or part of the system that we've created for worker safety and health is discussed. This is a step-up method, moving from examples to groupings. The groupings, however, allow us to see the problems that we've created and put in place in each part of the system. This approach also allows us to identify how to remove or minimize these problems or their associated impacts. Finally, the book closes with an overview, which allows us to look at how each part impacts on the whole system.

This is not an all-inclusive look at the worker safety and health system. To do this, volumes would be required and with all the changes happening today, would most likely be obsolete long before they were published. Besides, the details would only provide historical perspective and be of little use to us in our understanding of the system.

The identified problems are not presented in excruciating detail. We'll touch on the major ones. You may see others. That's good: adding yours will increase the impact of the material presented and personalize the problem-solving.

Writing this book was an adventure in every sense of the word. Too often we just curse at a problem and then move on—like encountering a pothole in Omaha on a drive across the United States. "Damn, why don't "they" fix it!" Moments later, the problem is forgotten as we continue our journey. Researching the material for this book required that I get to know the entire route very well. I had to become intimately aware of each turn, each blind corner, each traffic light and stop sign, each intersection, each route division and interchange, each city and town—and each pothole too. But, more than knowing that each existed, I had to stop waiting for "them" to correct deficiencies. Really, we safety and health professionals are the custodians of this system for worker safety and health. Each of us needs to get to know it much better and take ownership of fixing the problems.

Awareness leads to knowledge, knowledge to intimacy, intimacy to ownership. Too often we are satisfied to be just casually aware of the parts of the system that directly impact our jobs. We must immerse ourselves in the system, become knowledgeable, allow our knowledge to sweep us to where we can become intimate with the system. Only through intimacy with the system can we stop looking away each time we hit a pothole and demanding that others fix it. Intimacy is linked to ownership and, through ownership, we take responsibility for finding the solution.

This book is divided so that each significant historical and current issue that impacts worker safety and health could be looked at separately. Each issue has differing impacts. None is heading in the same direction. Some appear static while others are dynamic. One point is irrefutable. The complexity of the entire system for worker safety and health continues to deepen. This has been a result of continual regulation, politics, court or review system interpretations, the changing face of American business, and our stubbornness to adopt new ideas and ways of thinking. Within each chapter we will strip apart each issue looking at history or intent where important. We will candidly talk about the problems that exist and impede advancement of worker safety and health and offer ideas on how each can be "fixed." Are all the possible solutions noted here? No. The intent of this book is not to be the definitive "how to" manual for correcting the problems. It is to focus our attention on worker safety and health in America and what system aspects hurt it. By offering my views on how to fix the various problems with the system, it is my hope that an opening volley be sent across the bow of the ship representing worker safety and health in America. The ship is slowing sinking by its own weight and inefficiency. But undaunted, it plows forward in the seas of complexity without direction or goal. As you read this book, you may agree or disagree. It really doesn't matter. It only matters that you also take aim and together we might have an impact at turning the ship.

1

FROM CARNAGE TO CONFUSION

Why is it that we never seem to perfect the safety and health system? I wish I could say that we try, and fail. But in fact, a lot of the time we don't even try. It's not a failure of American industry or government alone. It's a failure by management, workers, organized labor, government, and those of us in the safety and health professions. It's a failure of an entire system. Too many times, we choose sides and start throwing rocks at each other. Labor says that management is to blame for poor worker safety and health. Management faults the recalcitrant unions and OSHA or careless workers with poor attitudes. Safety and health professionals claim lack of respect for the profession, uncaring management and inadequate staffing levels. OSHA blames management and lack of funding. The fact is, they're all right, but they're all wrong too. It isn't a simple puzzle, but one fact is irrefutable. When worker safety and health in America are compared with that of other countries such as Germany, Sweden, and Japan, America always comes up second-best. Why is that? How do we start to sort out the pieces of this puzzle? As with any puzzle, you start by opening the box, and that's a very good place for us to begin also. Let's start with history—how we got to where we are today.

The concerns for worker safety and health aren't twentieth century issues. And, obviously, they aren't just American issues. Worker safety and health concerns have been around for thousands of years, even before the building of the great pyramids. History provides us with many accounts of worker safety and health issues. Worker safety responsibilities were even discussed in the Bible. Ancient Greeks like Hippocrates wrote of worker safety and health issues. Health effects from exposures to substances such as lead and mercury have been described for thousands of years.

In America, however, these issues didn't get much attention until early in the twentieth century. In the late eighteen hundreds, the industrial revolution struck our continent with hurricane-like force. The expansion of our industrial base blossomed overnight. Factories sprang up from Maine to the deep south, and up and down our waterways. The Midwest industrial belt grew rapidly. Not much thought was given to environmental issues and pollution at that time, nor to the safety and health of workers. However, no one can argue that we didn't know the issues that were involved there.

THE INDUSTRIAL REVOLUTION

The industrial revolution was good for our country from an economic viewpoint. Overnight, we became an economic world leader. Rich in natural resources, we became an exporting giant. No longer did we mainly export agricultural products. A boom occurred in the export of finished and semi-finished goods. Energy became even more important to support our expanding industrial base. To meet the need, more and more coal mines were opened. Coal mining towns flourished. There was money there and workers were all too anxious to suffer through the horrible working conditions in the mines to get some of those wages. In industry, things weren't any better. Safety and health wise, working conditions were subhuman. Known hazards such as rotating gears and fast-moving drive belts were left unguarded. Molten metal and toxic atmospheres were unchecked in the workplaces. Workers were exposed to all sorts of health hazards including dust, toxic chemicals and extreme heat. But again, there was a lot of money to be made and workers were anxious to trade income for the known hazards, injuries, and death.

One of the cities that grew up around an expanding industrial base was Pittsburgh, Pennsylvania. By the early 1900s, it was a booming industrial city, one of the largest because of its plentiful natural resources and its good transportation routes. It had all the hallmarks of a booming industrial city. The air was black and had many foul odors. The water was polluted. Particles fell constantly from the sky. The grime was in everything, in workplaces, in homes, in the food its citizens ate, on their children. Daily, workers would trudge off to the factories where even greater hazards awaited. They worked long hours for low wages in conditions we wouldn't subject prisoners of war to. Gloom not only described working conditions, it also described the workers' existence and their hope for the future. They were trapped by a system that neither appreciated them, nor took any responsibility for making things better.

Worker safety and health in both industry and mining were not considered a responsibility of the system, nor of government. Safety and health were responsibilities of the worker. The philosophy at that time was that there were hazards inherent in working in industry or in the mines. It was an accepted fact that hazards existed, period! Therefore, it was the responsibility of the worker to watch out for these known hazards and, as much as possible, to avoid them. This wasn't just the philosophy of industry, mining or management. It was prevalent throughout the entire system, including management, workers, and government. After all, the worker had a choice: to go to work and watch out for the hazards, or to stay at home and starve.

THE PITTSBURGH STUDY

A curious thing caught the eye of some researchers at that time. Was it the horrible working conditions in the Pittsburgh factories? Was it the ever-present black sky and raining particulates? No, it was the carnage, the body count that was coming out of the factories. In other words, it wasn't the disease that caught their attention, or even a symptom of the disease. It was the result. This speaks very eloquently about how desensitized the entire country had become to the system. Blinded to the obvious, the researchers focused on the result—the body count, the crippled, the maimed. The result was a year-long research study called

"The Pittsburgh Study." Between July 1906 and June 1907, the researchers counted the "major" work-related injuries and all fatalities in the Pittsburgh area. As part of the study, they kept track of fatalities by noting them on a calendar that hung on a wall. This particular result of the study was so attention-getting that it became known as "The Allegheny County Death Calendar." Almost every day, one or more "X's" (which signified that a work-related death had occurred) were added to the calendar. Over that year, there were 500 "X's" placed on the calendar. In their data that year, the researchers also counted "45 one-legged men," "100 cripples," "45 with useless arms," "30 armless," "20 with one hand," "60 with maimed hand," "70 with one eye." All in all, the published study said, hazards in the factories resulted in "500 human wrecks"; this was, of course, in addition to the 500 who lost their lives. Needless to say, the results of the Pittsburgh Study received a lot of attention.

BIRTH OF WORKER SAFETY AND HEALTH CONCERNS

Most experts credit the Pittsburgh Study as the opening volley in the war for worker safety and health in America. The Pittsburgh Study was interesting data and for the first time the carnage was documented, but it was only one point in the data. American business doesn't react to one datum point. We tend to be very conservative. Most business decisions need a lot more data before a trend is verified, and certainly before action is warranted or decisions made. No, it wasn't the Pittsburgh Study that brought about the safety and health revolution in America. There was yet another factor that caused the change in business' attitude concerning worker safety and health. As with most things that catch the attention of business, it was monetary.

During that period, if a worker was injured or killed at work, the only way of obtaining compensation or benefits from the employer was to sue. Keep in mind that around the turn of the twentieth century, there wasn't the density of attorneys that exist today. Consequently, most worker suits were pursued without legal counsel. Businesses, however, had ample access to legal talent. Not that there were a lot of suits, since there were too few judges for the cases presented and cases could take years before coming to trial.

As with any tort litigation, for a worker to obtain a favorable decision, he or she had to prove that the employer was negligent and, thereby, was responsible for the injury or death. In a system that openly accepted that safety and health was the responsibility of the worker to "work safely," that was not an easy thing to prove. The deck was also stacked against the worker who seldom had legal counsel. The injured worker was probably unemployed at the time or, if deceased, his family had lost their source of income and did not have access to good legal services. They would be faced with a very difficult choice: Would they put food on the table today with the little money they had, or spend it on an attorney and risk losing what they had? So, let's just say that most workers couldn't afford legal help.

A look at court records of that time show that worker court challenges against employers were rare. There was, however, a definite upward trend in litigation. Workers were beginning to question the long-held belief that their safety and health was exclusively their responsibility. They were beginning to feel that the employer had some level of responsibility for providing a safer and healthier workplace. The success rate of worker litigation against their employers for injuries or death wasn't very impressive, but the availability of affordable, qualified counsel to workers was improving.

So, if there weren't a lot of suits brought about by injured workers or their families and the success rates weren't high, what possible monetary reasons would there have been for business to change their attitudes concerning the responsibility for worker safety and health? There were plenty.

First of all, retaining attorneys was not cheap for business. Lawyers have never worked for little or nothing. Second, litigation tied up important people, including the owner or top executive of the business. Preparing for defense was also a disruption to productivity in the workplace. Third, when one of these grizzly accidents happened, the shock waves that went through the work force also caused ripples in production, big ripples. Management of the remaining work crew was much harder following such an accident. Fourth, performing investigations long after an accident for defense of a suit was expensive and took a lot of time. Also, important information that supported the case of the business was forgotten or not discovered. Fifth, in 1908, the federal government enacted a workers' compensation law that covered federal workers only. Sixth, there was *Brown versus the Long Island Railway*. This was a landmark case in which the Supreme Court concluded that employers had a responsibility for the safety of their employees. And seventh, business leaders have never been stupid. They have always been able to see obvious trends and calculate potential future costs should those trends continue. In short, a shift in attitude about the responsibility for worker safety and health away from workers and more toward business was a good business decision.

So, those who believe that businesses began to take worker safety and health issues more seriously for altruistic reasons following the publication of the Pittsburgh Study are mistaken. The change was driven almost totally by the balance sheet. It was a simple business decision based on monetary reasons alone. It was, in fact, a good business decision.

WORKERS' COMPENSATION BEGINS

As anticipated, the laws of the land began to follow the federal example. In state after state, workers' compensation laws began to be passed, beginning in 1915. But it wasn't until 1948 that all states had workers' compensation statutes on the books. It's important to note that workers, unions and businesses alike supported these workers' compensation laws. Why? From the workers' perspective, it eliminated the necessity to prove negligence. All they had to do was to prove work-relation. That was much easier and most of the time it sidestepped the need to get an attorney or go to court. It became a paperwork and verification system. Workers' compensation laws also provided a known amount of compensation. That removed a lot of the risk associated with tort litigation. From the employers' vantage point, workers' compensation laws were less costly. Using lawyers would not be the norm; that alone saved a lot of money. Production disruptions would also be greatly minimized. The laws also removed the uncertainty as to how much an employee or his family would be awarded by the court. So, from a business perspective, it was a more dependable and inexpensive way to do business. From both worker and business perspectives, workers' compensation laws made sense.

SAFETY AND HEALTH PROGRAMS BEGIN IN BUSINESS

An Economic Decision

In response, industries began safety programs. Again, it just made good business sense. Now that the costs of worker injuries, illnesses and fatalities could be easily calculated within a particular state's workers' compensation program, it stood to reason that injury prevention could reduce those costs to the business.

As an additional confounder, American business was and is organized in a "standardized" form with a line and staff structure of those who add value and those who are "costs of doing business" (called overhead.) This new cost-avoidance measure, safety and health, had to fit into the existing organization somewhere. Safety and health, as a cost avoidance measure, became an additional load to the overhead.

Because the creation of safety and health programs was a business decision that was based on economics and had to be absorbed into an existing organizational structure, the results were predictable. There were seven of them.

First, being economics-based meant that the preventive effort had to be economically justifiable. It is not a good business decision to spend $50,000 in order to avoid a potential cost of $1,000. This reality affected not only the correction of hazards in the workplace, but it also dictated the size of the safety and health staff and how much they were paid. It doesn't take an accounting genius to see the outcome. A "cheaper is better" mentality became prevalent. Lengthy discussions concerning the cost if a hazard was *not* corrected were inevitable. It became a balance sheet discussion. After all, workers' compensation had fixed the prices for fingers, toes, arms, legs, and lives.

Another fallout of this economic justification thought process was that safety and health staff were selected from those who were "available," and usually meant workers who were idle, no longer needed, eliminated by improvements, or injured. This agreed with two strongly held beliefs in business: safety was easy, and convincing spokespeople would make workers work safer. In other words, business at that time believed that safety was merely correcting obvious hazards and reacting to accidents, so a brain surgeon wasn't needed to do the job. Also, the foundry man who had a serious injury made a good "safety guy" because he was a walking testimonial to workers, illustrating what could happen if they didn't work safely. So, it was more important for the "safety guy" to speak good "worker-ese" and be an example than it was for him to be skilled in this new profession. After all, business thought the workers probably couldn't understand cost-benefit evaluations. And besides, that was a management decision process anyway.

Dollar Value of Worker Safety

Third, because safety was economically based, there was no way to avoid sending a mixed message to the workers. This message is still misunderstood and miscommunicated to this day. It goes something like this: "We, management, are concerned about your (worker) safety and health. We care about you." The unsaid message, the message that workers *heard*, however, went more like this: "Because we are running a business, what we do to improve worker safety and health has to be economically supportable." Or worded more strongly, "Your life is worth so many dollars. Your arm is only worth this much, etc. Correcting the hazard depends on what cost it will avoid." But in reality, it was thought, "If you can't work safely, your life or arm probably aren't worth that much." Management did not send such a message to workers on purpose;

in fact, I seriously doubt if they even knew they were sending it. But the workers were receiving it loud and clear! Polarization of management and workers was the inevitable outcome. The safety and health guy was stuck in the middle, a friend neither to management nor to the workers.

Fourth, preventing or reducing costs has always been a "soft-index." It isn't like saving $10,000 when you buy a cheaper raw material or machine part. In that case, you know how much you have spent, and more importantly, what you have saved by your wise purchasing. Injury or illness prevention is much different. If injuries are reduced or avoided, how does the business know that the result is due to the preventive efforts (balanced by the overhead costs), just an industry-wide trend, luck, or safer workers? On the other side of the coin, when accidents increase, is the increase real or just a baseline miscalculation, a result of bad luck, substandard safety effort, or just careless workers? Consequently, safety programs in business have fought continuing battles to prove the economic viability of their programs. Also, in a "soft index" measurement system, how do you know when you are doing a good job (when you are winning the battle) or a bad job (when you are losing it)? You can't put on a balance sheet an amount for something that didn't happen.

Fifth, economically driven decisions have extremely short memories. In other words, "I know what you did for me last year; what are you doing for me now?" Thus, safety and health has been a "boom or bust" function. When safety is perceived as needing help, the effort "booms." However, when worker safety is doing okay, the safety guy gets caught in the "bust."

Safety as a Management Program

Sixth, because business accepted safety and health as a preventive responsibility, safety and health became business' program, management's program. Perhaps it is best described by a typical child's threat during a neighborhood ball game: "It's my ball, so I'll decide if and when I play and I'll make or bend the rules as I wish. If you don't like it, I'll take my ball and go home!" This attitude excluded workers' input from any part of safety decisions. Safety decisions merely flowed out of managements' offices downhill to the worker. Input equals importance. If workers had no input, obviously they weren't important in the safety effort. They were only pawns, another factor contributing to management/labor.

And seventh, safety had to be fit into the business structure, just like the other overhead functions. Once incorporated, the perception of its function was cast in the same way the other functions were. For example, who is responsible for engineering functions? Engineering is. Who is responsible for purchasing? The purchasing department is. Who is responsible for accounting? The accounting department is. Who is responsible for the safety and health of workers? That's right—the safety and health department or person is. Like engineering, worker safety and health was not considered a line management responsibility; it belonged to the safety program. This responsibility without direct authority has long been a stumbling block to worker safety and health.

STANDARDIZING THE PRACTICE

Safety Organizations

In any event, the safety and health movement had begun in American business. This left two gigantic questions which had to be answered. "What is safe and healthful and what isn't?" And secondly, "What safety and health performance is good and what is bad?" They are both important questions to any business; however, they are also questions that are too easily determined by perspective, and differing perspectives in business cannot be managed. There has to be a consensus, a standard, a common reference point.

As a result, organizations, associations, etc. were formed in order to develop recommended guidelines for business. Among these were the American National Standards Institute (ANSI), Underwriters Laboratories (UL), the American Conference of Governmental Industrial Hygienists (ACGIH), the American Society of Mechanical Engineers (ASME), and the Compressed Gas Association (CGA), to name a few. These important standard-recommending organizations and associations began an arduous and long-term effort to provide answers for the question, "What is safe and healthful and what isn't?" Because this was such a monstrous undertaking, it started slowly. As a result, recommended standards never came fast enough to meet the demands of the new technologies and applications that were being used in American industry and mining.

Developing Safety Standards

This created a large level of uncertainty within safety and health efforts. Was this a problem? In a system where safety and health measures were economically-based, it certainly was. Having no recommended standard in place left the door wide open to questions such as: "How do you know it's a hazard?" "How many people have been injured by it?" "How do you know that it will injure or kill someone?" "Why should we fix it if we can't agree on whether or not it needs to be fixed or how to solve it?"

The question of, "What is safe and healthful and what isn't" became problematic. Without a consensus or a common reference point this question was answered solely by perspective. From management's perspective, because safety and health was an economic decision, the answer was one of economic justification and cost-benefit. But from the workers' perspective, the answer was extremely personal—it concerned their fingers, their arms, their legs, and their lives. One perspective is cold and logical; the other is personal and emotional. Neither provided common ground between management and workers. Another polarizing hurdle was being built.

An expanded version of this question has troubled both workers and safety and health professionals for years, and derives from an extremely aggressive business posture. The expanded position would translate this way: "It is only recommended that I do this. Why should I? Who is going to make me?" This extreme position comes from many sources, among them Theory X management. That is, since it was not management's idea, it obviously had little value. Management ego was another source. But probably the largest impactor was the business culture that existed: "We are management and you are labor." "I manage and think. The success of this business is my job. You make the product, mine coal, drive the truck, jackhammer the street, etc. (select one). You are not paid to think, only to do what I say." This culture was evident in other ways, including executive lunchrooms, reserved parking places, management clubs, and the pay system. This expanded position of management didn't place a polarizing hurdle between workers and management; it *was* the divider, an accepted and common separation of castes.

Safety and Health Performance

The second question, "What safety and health performance is good and what is bad?" also needed to be answered. The answer, however, was easier—develop a common statistical index. The injury rate was invented and the National Safety Council formed. The agreed upon index, the injury rate, was standardized on one hundred workers for one year. Using a rounded number of work-hours (2,000 for each worker), a hundred-worker base could easily be viewed as a percentage of any work force whether it was 25 or 100,000. Thus, an injury rate of five meant five injuries out of 100 workers in one year or five percent of any number of workers per year. It was a convenient index and one that could be easily used for comparison.

The National Safety Council was formed by industry members to compile statistics on injuries and fatalities across American business. But the quality of the comparison depended greatly on how many businesses belonged to the Council and provided injury and fatality statistics for the Council to analyze. It began slowly at first. A lot of businesses considered their injury and fatality information to be extremely confidential. But, over time, that belief changed, again driven by business concerns about how to manage a safety and health function without a reference point. The National Safety Council has for many years provided this valuable service to business.

A MATTER OF LAW

The OSH Act

The Williams-Steiger Occupational Safety and Health Act, otherwise known as Public Law 91-596, was signed into law by President Nixon in December 1970. The federal law created four bodies: the Occupational Safety and Health Administration (OSHA) within the Department of Labor, the National Institute for Occupational Safety and Health (NIOSH) within the Department of Health, Education and Welfare, the Occupational Safety and Health Review Commission within the Department of Justice, and an advisory committee for safety and health. OSHA, of course, was the enforcement arm of the OSH Act. NIOSH was formed to provide research and education in occupational safety and health. The Review Commission was the judicial arm that was to resolve contests to OSHA citations and actions. The advisory committee was to provide direction for OSHA and NIOSH.

Workers had strongly supported passage of the OSH Act while business, for the most part, had resisted its passage. This was a "head in the sand" position. With the polarization between labor and management, the different positions of business and labor were not surprising. Workers, of course, felt that, without the force of law, safety and health hazards were too negotiable in an economic-based occupational safety and health system. Business, however, because of their very different perspective, felt that having safety and health "nuts and bolts" defined by law was unnecessary. They felt everything was working well as it was. In support of their position, they pointed out that worker injury and fatality statistics had continually fallen in the non-mandatory environment. Oddly enough, the opinions of safety and health professional organizations were mixed concerning passage of the Act. Some position papers were sent to Congress, but at best, they were watered-down.

Consensus Standards

With passage of the OSH Act, OSHA was given a very tight time frame in which to develop standards that would have the force of law behind them. The OSH Act provided some direction for this effort in that it allowed OSHA to draw from current federal statutes and from "consensus" standards. Consensus standards were defined as those which were developed through participation from all sectors, including government, academia, industry, and labor. The American National Standards Institute (ANSI) and other consensus organizations were obvious resources for this effort. Within this very tight time frame, however, two significant problems emerged. First, because of the extreme detail included in the many consensus standards, it was impractical to make all of the recommendations law. And, second, the organization that had the most thorough listing of health exposure recommendations was the American Conference of Governmental Industrial Hygienists (ACGIH). But it was not a consensus organization, so its limits could not be used.

To counter the first problem, too much detail in the recommendations, massive cut-and-paste efforts were quickly undertaken. Due to the amount of material and the mixed expertise of those who did the cut-and-paste job, the outcome was regulations that were significantly flawed. For example, some standards contained too much detail. The requirement for split toilet seats was a good example. And some standards omitted significant parts of the recommended guidelines, such as knowledgeable program administration for respiratory protection programs. The authors of the recommended guideline for respiratory protection had thought this point so significant that they set it aside from the requirements for a minimally acceptable program and highlighted it. Because it was separated, the OSHA lawmakers missed it when they hurriedly pulled parts of the consensus standard into the regulations.

The second problem, that the ACGIH was not a consensus organization, was quickly and easily resolved. Because the act allowed usage of existing federal statutes, the ACGIH exposure guidelines were taken from the Walsh-Healy Act which had been passed earlier to protect government workers. The Walsh-Healy Act did not have these same "consensus" organization requirements because it only applied to government workers. Using the exposure limits from the Walsh-Healy Act dated the limits by a couple of years, however, as the ACGIH recommended limits are updated annually. At that time, this was a minor problem.

From Guidelines to Regulations

Other than the obvious problems brought about by quickly putting together cut-and-paste regulations, there was another major detractor to adopting recommended guidelines as legal requirements. When these guidelines were developed, they were never intended to be law. Therefore, the recommendations were either skeletal, due to a lack of expertise, or the length of development time, overly detailed and expansive, which was a more significant problem. This was done consciously by the guideline writers because they knew that the purpose was to be a general guideline for steering decisions, not dictating decisions. But making these "intended guidelines" "mandatory" and "overly detailed" caused a lot of animosity in industry toward the OSHA regulations. This lead to inescapable frustration with compliance efforts and set the stage for larger problems of the continued belligerence by business, and disagreement over the formats that would be used on future "troubled" standards.

There were two other very significant flaws within the OSH Act and in how the regulatory system was developed and implemented. The first flaw was the standards promulgation process, and the second was OSHA staffing levels. The wording of the standards promulgation portion of the OSH Act received a lot of comments, discussion, and amendments prior to passage. The fears lay on two sides of the issue. Some were greatly concerned that new standards should not be promulgated at the "drop of a hat." The OSH Act needed to require an adequate review and comment period. This was a very conservative approach to new standards development. The other side was deeply concerned that new standards would never see the light of day, at least not easily nor in a timely manner, if the promulgation process was too cumbersome and lengthy. Anyone familiar with the Act knows that the overly-conservative side won and that the worst fears of those who were concerned with viability in standards development were ultimately realized. The fact is, the standard promulgation requirements of the OSH Act cannot keep up with the holes that existed in the original OSHA regulations nor with the expanding hazards, technology, and research findings that continue to evolve. It's a shackled elephant trying to race a swallow. Promulgation history has painfully borne out this fact.

OSHA Staffing and Continuation

The other flaw was in OSHA staffing. In short, there were never enough inspectors to do the job the OSH Act required. So, not only was industry saddled with a "can't get there from here" situation of trying to comply with the regulation's detail, but OSHA was not able to do the job the Act required because of inadequate staffing. For both, it was "Catch-22."

There are those who argued and continue to argue about the need for OSHA and mandatory occupational safety and health regulations. There are still others who continue to argue against the limitations placed on OSHA and the "soft" parts of the Act and regulations. Both are emotional positions. At times, we find ourselves on one side of the argument and at others, on the reverse side. Oddly enough, most of the time we find ourselves on both sides at once. After all, it makes little sense to have out-of-date regulations that cannot possibly keep up with changes in the workplace due to the Act's baggage nor to have a staff so limited that it cannot adequately survey and/or enforce those regulations. It's a paradox at best. Because of that, it's pointless to discuss the issue of OSHA's existence. It exists, and will most likely continue to exist. As it is part of the federal government, it would be an oddity if it made perfect sense or ran efficiently.

WHERE WE ARE TODAY

Worker safety and health has changed a great deal and, at the same time, very little from the early 1900's. Let me clarify this seemingly contradictory statement this way. A lot has happened. We know a lot more about worker safety and health today than we did then. Obviously, America's workplaces are much safer and healthier today. Occupational safety and health has gone from a "worker beware" issue to one of law with detailed regulations. There are many more advocates for worker safety and health today in government, academia, labor, and management and business. Occupational safety and health professions exist where there were none in the early 1900s. Today the impediments to worker safety and health are also more diverse and numerous. They are indeed a product of the system that created the need for safety and health and the way we pursue it. The system is what hasn't changed. While it's a lot more complex, it remains the same. More importantly, it *still doesn't work*!

There is an important perspective that we can gain from a frog. Scientific experiments have taught us that if you drop a frog into scalding water, he will quickly jump out and avoid injury. If, however, you take the same frog, place it in water at a comfortable temperature and begin to slowly heat the water, something different happens. The frog remains in the water long past the point at which injury will occur and it cooks. How is this interesting but twisted information about frogs important to our perspective about worker safety and health? In a lot of ways, we are like that frog. If we magically awakened in this time period and were plunged into the worker safety and health system as it currently exists, we wouldn't stand for it. Like that frog, we would immediately recognize that it was inhospitable. We could at that point choose to simply get out or do something to fix the problems in the system. But that isn't the case. Like the frog who slowly and happily cooks because it doesn't recognize that it has a serious problem, we have grown up in and with the system and are comfortable with the way it "feels." We have lost our perspective. Unable to step back and evaluate what we have created and nurtured, we do not recognize that the system doesn't work and is, frankly, inhospitable to worker safety and health. The system fails in what it was created to protect and advance and we don't see it. Like the frog in the slowly heated water, worker safety and health is beginning to cook. We've lost our perspective to see from the "outside" and can only see from our "inside" position.

"Workplaces are safer and healthier today," you might argue from your "inside" perspective; what's wrong with that? This point too, is not without argument. We argue passionately about the validity of the statistics, about which indicators are more important, and if worker safety and health is truly better today than just twenty or thirty years ago. For our purpose, let's say that it has improved. If you are content to have American business run second, third, fourth, or worse in the world market, then nothing is wrong with it. But, if you fully expect and demand that American business be the best, how can we accept second-rate worker safety and health? Just as America bears full responsibility for the slippage of our business position in the world economy, we have the ability to turn it around. And, for the sake of our children and their children, we must do so. On that same note, through the years, we created the system which impedes worker safety and health. Over the years, we have nurtured it. We have allowed it to exist in plain sight. Why don't we demand that the system be fixed so that America's worker safety and health can also be the best in the world?

No level of worker injury, illness or death is acceptable. If you believe differently, perhaps you should put this book away. To change culture, there must be a clear vision and mass commitment to that new vision. The system is part of our culture. And if no worker injuries, illnesses and deaths is the only acceptable vision, we must change that part of our culture. Of course, no one person can fix the system alone. It must be a concerted effort. But the effort begins with identifying the problems and beginning the discussion about how we solve them.

So, in a way you might say that in America worker safety and health has evolved. It has indeed evolved. It has evolved from carnage to where it is today—confusion, inefficient, ineffective confusion.

2

SAFETY AND HEALTH IN A LINE ORGANIZATION WORLD

WHO MANAGES WORKER SAFETY AND HEALTH?

A young safety and health practitioner had spent a few years gaining experience when THE opportunity came. A medium-sized manufacturing facility was trying to find a young, aggressive safety and health specialist to replace another who had left the company. The facility had some significant challenges. They knew that. The previous person had left for a "better job" almost six months ago. The company had been looking since the previous person had announced his departure, but they hadn't gotten around to filling the position. Something always came up and interrupted the process. The young safety and health practitioner was not offered a great salary, but it was sufficient. It was the opportunity that he looked forward to. Six months later, he too had left the firm for a "better job." Somehow, he and the company never found common ground. He and the company's management had concluded that he was ineffective. It was a mutually agreed upon separation.

Why did this happen? Anyone familiar with the practice of safety and health knows that this is not an isolated occurrence. Why did this young aggressive specialist not find the opportunity he expected and/or why did the experience come to an "ineffective" conclusion?

19

The answer is complex. As this occurrence is also common, so are its causes. The causes were bi-directional. The young specialist was ill-prepared for such a challenge. Schools don't prepare students with the practical knowledge they need to accept such a position so soon after graduation. And he hadn't learned enough from his previous experience. Schools teach specific knowledge. They don't teach how to use it or, more importantly, how to apply it. The specialist also lacked knowledge and experience in dealing with a line-management culture. In other words, he didn't speak their language, know how to communicate issues, organize efforts or sell what needed to be done. The company was too traditional in its concepts of line-management responsibilities. In American business, this is not considered bad. Because of this, it's common. It represents another part of the system for worker safety and health which doesn't work. This disconnection between the specialty of safety and health and the traditional line-management world is important to explore. No matter how you slice it, excellence in worker safety and health requires that both specialty and line management work in orchestrated harmony. So let's talk about safety and health in our line-management world.

It makes sense that when trying to understand the problems associated with the system for worker safety and health, it is important to begin the search at work itself. This, after all, is the focal point, the trigger mechanism for creation of all the other parts of the system. The other parts we will discuss in this book were created as support functions for or as results of breakdowns within the work environment. Why were worker safety and health laws created? In response to a perceived inattention to safety and health measures in America's workplaces. Why was the workers' compensation system born? Because of breakdowns and gridlock in resolving workplace injury and illness issues. Why were safety and health professional societies and associations begun? As support for and to provide information to those practicing worker safety and health in the workplace. Why were unions started? To resolve critical issues in the workplace that management wasn't addressing. Every part of the system for worker safety and health was created to resolve or address breakdowns in the workplace. So to begin our look into worker safety and health, the best place to start is the workplace.

The practice of worker safety and health in American business has traditionally fallen into one of two modes: either the responsibility is designated or it is not. Companies where this responsibility is not designated follow ancient management thought that worker safety and health is just common sense and that workers being careful (working safely) is a major part of the equation. In this mode, worker safety and health is assigned to no one in the organization, other than the worker. A slogan campaign or a witch hunt usually follows an accident. In these workplaces, injuries are found to be caused by inattentive workers, those with poor safety attitudes, whiners, or those out to "get" the company. Because of the financial constraints of small businesses and a lack of broad business operation knowledge, this typical structure has no one designated as responsible for worker safety and health. In larger businesses, however, it is more a function of management ignorance and the continuation of an antiquated, yet traditional business culture. These workplaces become more "sweat shops" than progressive workplaces. We cannot passively ignore these workplaces, though. It is estimated that about 70 percent of American workplaces employ less than 10 people. And workplaces with less than 20 workers, employ about 50 percent of American workers.

The other mode is where someone or some group is assigned the responsibility of worker safety and health. Unfortunately, designation of responsibility alone does not insure success. Why? Because the business environment and organization typically don't change. Traditional American business falls into the category known as a "line management world." Designation of specialized responsibilities that fall outside of this "line" thought process has the same problems. These problems are created by the way these line management systems work. So . . . how do they work?

THE LINE MANAGEMENT WORLD

The Organizational Pyramid

Traditionally, American industry has been organized into a line and staff structure. To a greater extent, this structure is also prevalent in the mining and non-manufacturing businesses and the service sectors. It's pretty standard. The line portion of the organization is a power-based chain, an authority structure. Typically in corporations, it begins at the Chairman and/or Chief Executive Officer (CEO), and goes, perhaps, to a Chief Operating Officer (COO) or president or presidents, and on down to various levels of vice-presidents, to general managers, to plant managers, to department managers, to supervisors, and finally to the workers. Spreading out at the bottom, it has been called the "Organizational Pyramid." The various levels may be more or less and the titles may be different, but an important point to note is that authority starts at the top and moves down through the line structure to the bottom. At each level, the amount and scope of authority is diminished.

Responsibility is shared throughout the line structure. At the top, responsibility is more holistic. At the supervisor level, it is more detailed and day-to-day. So, ultimate responsibility rests at the top but the performance depends on how well those lower in the structure perform their more detailed responsibilities. Authority is tied to the level of responsibility. In a line structure, the reporting pathway travels up the chain until it reaches the top. The worker reports and is responsible to the supervisor, the supervisor to the department manager, the department manager to the plant manager, etc., etc.

The line structure is based on direct descending power. The military in its organizational power is great at the top and decreases as the line structure descends. Each level of power is granted by those above. At each level the power is almost guaranteed. Front-line supervisors or foremen are the least powerful, department managers more so. Division management has even more power and as one advances higher up the line structure, he or she is rewarded by increasing power. The line structure is a power hierarchy.

Economics of Management

What is management's job? As Peter Drucker puts it, management's main job is to ensure economic success of the business[2]. Drucker, considered to be a management guru, talks about management's two aspects—the economics of business and people. After all, it is argued by Drucker and believed traditionally by American management that business is totally an economic entity. It survives or fails by economics, known as profit or loss. Everything depends only on the "bottom line." No exceptions. No arguments. Management is there to ensure the economic success of a business. This, you'll probably agree, is a rather cold perspective of management. No wonder American management has an image problem.

Traditional management thought separates management from labor. Managers are seen as the thinkers, those upon which the fortunes, both future and day-to-day depend. Workers are doers. This "we think, you do" belief is common in American business. It is an American business paradigm.

Further, management sees two important concepts, value and cost, very differently. At first glance, these two terms seem synonymous, but they aren't and the difference between them is important. Something is valuable if it directly serves the business' production or service generation. In other words, it enhances business ability to serve their customers. Costs, however, are merely detractors from the "bottom line." Often referred to as "the cost of doing business," management capitulates to their existence and openly argues about how much cost the business can endure or afford. However, there can never be enough values. They only add to the "bottom line." It's a credit versus a debit issue to line

---

[2] *The Practice of Management* by Peter Drucker

management. Value and cost are significantly different concepts in American business.

STAFF FUNCTIONS

Anyone familiar with business or organizational structure is aware that there exist many support functions. Drucker talks about these in his book, *The Frontiers of Management*. These support functions naturally fall outside this line structure. They fall outside because they provide specialized services across different levels of the organization. These have typically been called staff functions. They have included specialties such as accounting or controller (an oddly appropriate word choice), purchasing, marketing or customer services, computer services or management information systems (MIS), engineering, quality assurance or product assurance or quality control, shipping and traffic, human resources (a strange choice of words) or employee relations (an even stranger choice of words), environmental, security, medical, and safety and health or regulatory compliance (a traditional role-trap by name.)

These staff functions are normally flat in structure and are stuck into the line organization at reporting levels which make sense to the overall organization. For example, it is common to have staff functions at the plant, regional, business unit, and corporate levels. (This too is changing.) Each function is directly reported to the plant manager, the general manager, the vice president, and the president or CEO, respectively. The staff functions "liaison" with the line structure and "matrix" to similar functions that are higher up in the organization.

Unlike the line structure, authority levels vary greatly between staff functions. Nothing is guaranteed or automatically given. Some have a little and some have a lot, depending on the "power" of that function within the organization and/or the level to which they report. Usually, it's a politics thing. Those who are good politicians and get the ear of "the man" (or woman) at the top have high levels of power or authority. Those who play politics poorly or are not "well thought of" are on the outs and, consequently, have low authority and power. Within a staff structure, authority is based totally on power and politics. However, unlike the line world, in the staff structure, power is *not* guaranteed. Staff responsibility is generally limited to administration of their particular function or specialty. And that is tightly defined.

In smaller organizations or facilities, it is common for one person to have the responsibility for many staff functions. It is called "wearing many hats." In larger organizations, each function can require many people.

Theoretically, the size of a function's manpower is directly related to the size of their job. More realistically, however, it is more closely aligned with the function's power. And remember, in the staff structure, power is *not* guaranteed. It is given by line management.

SAFETY AND HEALTH AS A STAFF FUNCTION

It is a logical decision on business' part to place safety and health within the staff structure for many reasons. First, it is not directly involved with the manufacture of products or the services that the company provides. In other words, the safety and health function is not a direct contributor of income to the business. Second, because it is primarily involved in the administration of a function, it can provide support across many levels and many line groups within an organization or facility. So, there is a correlation to the line organization. It isn't a 1:1 ration, safety-to-line. Depending on need (or more appropriately, power), it is more a 1:500 or 1:2,000 ratio that is based on a perceived economic need by management. You can see the obvious opportunity for management to play with the ratios. And third, there is no convenient way of inserting safety and health expertise within the line structure. In such an effort, duplication and overlap is inevitable.

On the other hand, safety and health fit poorly into the staff structure for one gigantic reason, a reason that is common only to the safety and health function, not to any other staff area. Like production, the safety and health of workers is completely dependent upon direct authority over the day-to-day function of workers. In other words, each employee, as a part of his or her job, is required to work in a safe manner, take reasonable precautions, and work as trained or as instructed by guidance documentation. How can a staff-oriented safety and health function be responsible for the safety and health of workers without direct authority over their job assignments, the instructions given them, the procedures by which they work, the equipment they use, the training they receive, etc.? It can't be done! This is a direct disconnection in worker safety and health within 95 percent or more of American workplaces. And it flows from the way "business has always been done," the traditional American workplace.

Let's look at areas of management responsibility. Where does the responsibility for meeting production levels lie within an organization? It lies directly with the supervisor, the department manager, the plant manager, etc. If production levels are not met, line management is accountable and expected to turn the situation around quickly. Production is a line management responsibility.

If there are problems with product or service quality within an organization, where does the responsibility for product or service quality lie and who will be "called on the carpet" to answer for it and correct it? The supervisor, the department manager, the plant manager, etc. will. Producing a quality product or service is a line management responsibility.

If a budget is out of control, spending is too high, whose responsibility is it to correct it? The supervisor, the department manager, or the plant manager is the one whose feet are held to the fire. Cost containment is a line management responsibility.

If a serious injury or illness occurs, the injury or illness rates are unacceptably high, or the workers' compensation costs are out of control, who is drawn and quartered? The safety and health guy! What? Hold line management responsible for worker safety and health? To do differently would be . . . well . . . un-American. After all, why is there a safety department anyway? The safety and health staff function exists to keep workers safe and healthy. Does this disconnection in common thinking seem strange to you? It isn't rare. This disconnection exists in 95 percent or more of America's workplaces. It's traditional.

"No," you might say, "it just isn't true where I work. It is definitely not true." There is a simple memory test that you can use to test this point, but you have to be honest. Think back. Remember a significant injury or illness. Or maybe there was a steep upward trend in injuries. Whom did the plant manager call in to "fix the problem?" Was it the area manager or the "safety guy?" Who was the brunt of the plant manager's pointed remarks, the department manager or the safety and health department? Who got "out-placed," demoted, or offered another "opportunity" when the problems didn't go away? We say, "Safety is Number One" and "Safety is management's responsibility." But answer for yourself. Are these realities in the vast majority of American workplaces?

SAFETY AND HEALTH POWER IN THE LINE ORGANIZATION

But the convoluted problems of worker safety and health within the line organization don't end there. What would be the most effective way of putting this? Let's try it this way: "Historically, in a line organizational business world, safety and health has had no pull!" Using more politically correct language, "Over the years, safety and health has had a limited power base in American business."In a staff orientation, all functions are basically thrown into a dog-eat-dog environment. Those who play politics, take better advantages of opportunities, and gain the ear of the "top dog" are rewarded with greater power. Power equals authority. Responsibility with authority (even if it may not be direct) is the only way that a staff function can succeed. On the other hand, those staff functions that do not have extensive power (granted by the highest member of management at that level) have little authority. Responsibility without authority or with minimal authority equals failure! Before that function even begins to try, it is doomed to fail.

There is a power relationship that is pertinent to our discussion. This power relationship describes the aspects and results of impactors to power in a non-line structure. This, of course, would include all normal staff functions in a business, including accounting, purchasing, human resources, engineering, safety and health, etc. It can be displayed in a power grid. There are two significant areas of power-deriving issues—the basis of the function and the ability of those in that function to be political. The top of the grid represents the four normal justifications for or basis of that non-staff function. The need for those functions can be determined by business stereotype, a documented need, a perceived need by upper management, and by a directive from higher up in the organization. The other power-deriving issue used in the grid is the ability of those who provide that function to work in the political arena (accustomed to being highly political, moderately political, or less political).

STAFF POSITION OR PROGRAM DETERMINED BY:

		Business Stereotype	Documented Need	Perceived Need	Organizational Directive
STAFF POLITICAL APTITUDE	Highly Political	**High**	**High**	**Moderate**	**Moderate**
	Moderately Political	**Moderate**	**Moderate**	**Low**	**Low**
	Less Political	**Moderate**	**Low**	**Very Low**	**Very Low**

Safety and health programs are not determined by business stereotype. Some places don't have safety and health programs nor people assigned to those responsibilities. On the other hand, human resources and accounting are business stereotypes. At best, the safety and health function can be justified by documented need. More commonly, it is placed in the organization or facility because of a need perceived by management or by corporate directive.

Traditionally, safety and health specialists are lousy when it comes to politics. There are exceptions, but they tend to be rare. In most safety and health functions, the specialists in management of those functions would be, at best, moderately political and, more often, less political.

Using these known aspects (ratings) of traditional safety and health programs and specialists, it can be easily seen why we don't seem to muster much power. In the best arrangement, we can only expect moderate power. More commonly, we can expect low power or very low power levels. In the dog-eat-dog environment of staff functions, we are fighting an uphill battle.

Low power is precisely the position that most safety and health functions in business find themselves in. Why else do you imagine that a commonly conveyed message of safety and health professional societies and associations is to "CYA," write it all down? Why else would the vast majority of safety and health functions in business be buried in another staff department such as environmental or the more common placement, human resources? Why else would safety and health coverage ratios in businesses commonly be between 1:500 to 1:2,500 or thinner? Why else would the safety and health function commonly be one of the "many hats" that someone (a singular person) in the organization wears? Why else would the pay scale for "safety engineers" in almost every instance be not just a little less, but significantly less than that of environmental engineers, electrical engineers, mechanical engineers, etc.? You want a short, to the point answer? Safety and health has no pull in American business!

WHY IS SAFETY AND HEALTH LIMITED?

"Safety Is Easy" Paradigm

Why is it that safety and health has no pull in American business? Is it because worker safety and health is not considered important by management? No. It is that way because of a business paradigm[3], history, traditional management thought, and also because of the original economic-based thinking in which worker safety and health began and continues. What is the business paradigm that guides this placement for the safety and health function? In business today, just as it has been for years, it is a strongly held belief that safety and health is a simple job that requires little expertise and training, and seldom utilizes complex problem-solving skills. I call it the "Safety Is Easy" paradigm. Want proof of its existence? Human resource departments in business use a rating scale to determine pay grades for exempt and nonexempt job positions. It's a common practice in business. No matter what company you look at or survey (look at your own) if you look at job ratings for safety professionals, they are *always* rated low. Lower than engineers, lower than environmental professionals, lower than almost any other specialized staff position with comparable education and experience requirements. The low marks given to the task analysist is irrefutable evidence of this business paradigm. The "Safety Is Easy" paradigm exists.

But before we make too many conclusions, let's be fair in our assessment. Could this "Safety is Easy" paradigm be true? Is the function of worker safety and health easy to do and be successful at? Let me answer that question with another. Have the issues and regulations concerning worker safety and health changed over the past fifty years, the past twenty, or the past ten years? Unless you've been in a coma, you would have to answer with a resounding YES. In fact, they haven't changed just a little, they've changed drastically. Look at the change in concern and business impact of such safety and health issues as carpal tunnel syndrome, stress-related illnesses, asbestos and other fiber exposures, electromagnetic frequencies, repetitive trauma, ergonomics, lasers, and robotics, just to name a few. The world has changed

---

[3] A paradigm is a set of internal rules and regulations that establish boundaries and tell us how to be successful within those boundaries [Kuhn and Barker]

drastically and so has the practice of safety and health. No. The "Safety Is Easy" paradigm cannot be true. There's no way it can be!

To paraphrase Drucker, this "Safety is Easy" paradigm could be rephrased to say that the expertise is less important than business experience. In his book, *The Frontiers of Management*, Drucker strongly suggests that staff personnel must have operational experience to be any good to the business. It does not necessarily equal that the practice of safety is easy. It only means to Drucker, and many others in business, that operational experience is more important than expertise. I'm sure that if Drucker were more aware of the complexities of safety and health program management, he would probably rephrase his statement to include at least a sentence on the importance of knowing what you are doing in order to protect the economics of the business. He, of course, did not. Many downplay the expertise and fall into the "Safety Is Easy" paradigm, just like Peter Drucker.

Limitation of Power

History has also fed the continuation of limiting the power base for worker safety and health. To a large extent, history reflects the business paradigm concerning the ease of doing safety, the "Safety Is Easy" paradigm, and its existence for so many years. Those of us that have been taught to recognize and challenge paradigms as a way toward total quality and continuous improvement recognize the patterns of speech and phrases that are commonly used when a paradigm is involved. We have learned to recognize these phrases because they expose paradigms that limit creative thinking, and usually get in the way of making things better. "That's the way it has always been." "That will never work." "It just doesn't work that way around here." All of these are blocking phrases that flag a paradigm that has existed for enough time to become part of that organization's culture. One of those phrases can best be used to describe how history has limited safety and health's power in business: "That's the way it has always been."

The original economic-based thinking that formed safety and health efforts in business also greatly limits its power. Safety and health was adopted as a business responsibility to control costs. Worker safety and health has always been thought of as a necessary overhead expense to avoid costs. It was a balance sheet item. Never was it thought of as a value-adding function that directly contributed to the income of the business. Did you notice the difference between those two thoughts? One is negative and the other positive. From a business perspective, the negative thought might best be phrased, "We have no choice, we have to have a safety and health function. The only question, of course, based on reasonable economic impact estimates, is the size of the safety and health effort and the cost of compliance. It must be economically justifiable." However, the positive thought would sound like this. "Worker safety and health is valuable to the organization because it increases our return on investment (ROI)." One's a "got to." The other is a "want to." This is a big difference in thinking.

IS SAFETY AND HEALTH A VALUE OR A COST?

How about it? Is worker safety and health a cost-avoidance, overhead charge requiring economic justification, or is it a value-adding contributor to income? That's not an easy question to answer. In some organizations, safety and health is truly a value-adding function, while in most others it is relegated to cost-avoidance. Which side of the fence, value-adding or cost-avoidance, worker safety and health finds itself on depends on two critical affecters: the environment in which worker safety and health is placed, and the manner in which the safety and health function is performed.

The environment in which worker safety and health is placed is a function of management. It is completely a function of management, not labor, not luck, not regulations, not culture . . . management! If management *chooses* to maintain an environment where safety and health is relegated to the historical role as a cost-avoider, that role is inescapable. If, however, management *chooses* to "create" an environment where worker safety and health efforts can add value to the income of the business, provided the other affecter is satisfied, then safety and health can become just that, a value-adding function of the business.

CREATING A VALUE-ADDING ENVIRONMENT FOR SAFETY AND HEALTH

Safety Is *Not* Easy

What's involved in creating that value-adding environment? First, there has to be a change in the business paradigm concerning the safety and health function, the "Safety is Easy" Paradigm. The fact is, injury and illness prevention is *not* a simple thing to do. It requires advanced skill and knowledge, continual problem-solving efforts, working closely with regulators, interpreting regulations and standards, actively being a liaison with the organization's personnel at all levels, advanced communication skills, application of creative training methods, keeping current with regulations and research, and administering a highly complex and multifaceted function. Just as business practices are very different from even twenty years ago, so is the practice of safety and health. Neither is simple or easy.

Sure, Drucker was right also. Safety and health practitioners must also have an excellent grasp of business and economics to feed their end of this value-adding environment. But this is a dovetail change that must occur in the professions. We'll talk about this later.

Safety Is Management's Responsibility

Second, the safety and health responsibility must be placed where it belongs, with management. This cannot be in word only. Truly, safety and health *must* be management's responsibility. Just as management is responsible for costs, production, and quality, it must be responsible and held accountable for the safety and health of workers. This responsibility must include active observation of employees, identifying and correcting hazards, training workers in how to work safely, providing the necessary safety equipment, regularly inspecting the workplace for hazards, providing clear work instructions and identification of hazards and precautions, and providing communication pathways where safety and health concerns can be addressed. The function of the staff safety and health function must be one of coordinating and standardizing the organization's prevention programs, keeping the necessary regulatory reports and programs, and providing a service to the line management's worker safety and health effort. In such a responsibility alignment, line management has responsibility *and* authority. This is the only way it works!

Worker Participation in Safety Programs

Third, with respect to workers and their role in safety and health, the perspective must be changed from a negative focus to a positive one. In other words, a worker's role in safety and health has commonly been reduced to just working safely. When an accident happens, it becomes a search for violations. With this old perspective, it is concluded that if a worker is injured, he or she must have violated a rule. This is a negative approach to worker safety and health. This perspective must be turned 180 degrees. A positive approach focuses on participation of workers in safety and health activities and programs. For example, worker participation could include presenting training or awareness subjects in workplace safety meetings, creating safety and health awareness posters, identifying *and* correcting hazards or unsafe actions of fellow workers, serving on problem-solving teams, and reviewing and updating safe work directions. A worker safety and health program that is focused on encouraging and rewarding participation in safety and health promotion activities is positive and encourages workers to get involved. Programs that are closed to workers and emphasize a search for violations once something goes wrong are negative, discourage worker participation, polarize management and workers, poison improvement in worker safety and health, and force workers into defensive positions.

Thinking Beyond Economics

Fourth, management needs to stop playing games with the safety/health-to-worker coverage ratios as if it were totally economically based. In a value-adding organization, that correlation to economics is loose at best. The question shouldn't be "How many safety and health people can we get by with?" It should be, "How many do we need to support the hazard level of the business and to meet the service needs of the line structure?" If the worker safety and health function is to add value to an organization, it cannot be treated as a just cost. If you make a decision to convert process operations to computerized and pneumologic control, do you increase the engineering staff and expertise to build and maintain such a system? If you develop a new product or service line, do you bring in a person to market or administer that new area? These are not considered costs; they are considered necessities. Then does it make any sense for business and the regulatory community to change so radically and expect a totally economics-based safety and health coverage to provide a value-adding service?

Reporting Structure

Fifth, the safety and health coordinating function, the staff function, must *always* report directly to the top person in management. In Europe, this reporting structure is required by law. They felt so strongly about it that they wrote it into their worker safety and health laws. Why did they do this? It's really simple. They felt that it was the only way to assure that the safety function had the ear of the "top dog." In America, the reporting structure for the safety and health function is a mixed bag. There are no regulations that address it. So it is very common to have the staff function buried in human resources, environmental, engineering, etc. That "deep burial" takes important communication pathways and opportunities away, and greatly reduces the "clout" of the safety and health position. That makes "selling up the chain" a normal process before important safety and health issues can see the light of day. This "burial" is but common in most businesses and is highly inefficient. Look at the five holistic responsibilities of management: production, quality, costs, morale and safety. What is the reporting structure of the line and staff areas that oversee these important responsibilities? There is always a production manager. There is always a quality manager who 99 percent of the time reports to the highest management level. There is always a controller who oversees costs. That position also always reports to the top. There is always a human resources manager who oversees worker morale issues. Like the other positions, human resources also report to the top manager. Where is the overseer of the safety responsibility? Buried in the organization. Does this make any sense? It only makes sense if worker safety and health is of secondary importance to the business. If safety and health is to be in a position to add value to an organization and the function is to be equally important as a management responsibility, it *must* be a direct report to the head manager.

Changing the "Injury Rate" Paradigm

And sixth, business and management must challenge and change the "Injury Rate" paradigm. This paradigm says that the lower the injury and illness rates are, the better the worker safety and health program is *and* the safer and healthier workers are. Historically, this paradigm produces notable problems. First, it forces management to make unrealistic injury and illness objectives that are counter productive to the management effort. Rates are impacted by too many internal and external influences that simply cannot be managed. For example, what happened to injury and illness rates when carpal tunnel syndrome surgery came into vogue? If the rates go up, knee-jerk management results where a flurry of activities are "thrown" in the direction of the problem. That's not management. That is only Pavlovian response. Second, because the injury and illness rate is so hallowed, it causes good people to play terrible games with the records and numbers. If you want excellent examples of this, look at the multiple OSHA citations for recordkeeping violations that have cost up to the millions of dollars each. Third, when a serious injury results, management shoots the messenger, usually the safety and health guy or gal. The truth is, the injury or illness rate is more often a function of unmanageable internal and external influences. And in reality, injury and illness rates only bring about suspicion from OSHA and from workers that the books are being doctored and management is playing games. In response, OSHA is continually modifying the recordkeeping requirements and interpretations, and labor doesn't believe the numbers.

The only way to remove this "Injury Rate" paradigm is to begin to report everything—every injury and every illness. That, of course, would have four results. It would overload OSHA's system with data. OSHA would no longer have reason to be involved with the recordkeeping rules. Workers' suspicion of management playing with the books would go away. And management would finally realize that the injury and illness rates are really just numbers.

CHANGING THE WAY SAFETY AND HEALTH IS DONE

The other critical condition to whether or not worker safety and health can become a value-adding resource for business is how the worker safety and health function itself is accomplished. Has business changed over the past ten or twenty years? Of course it has. Business has had to change. Conducting business as usual would only mean failure. How then can we expect to run our safety and health programs as we have in the past? If management changes the organization's environment, as we've previously discussed, the stage is already set for changing the way safety and health functions are performed. In such a new environment, the safety and health function must also change. Personally, I'm partial to the total quality approach because it aligns the safety and health function with the way business is changing or will ultimately change. In a lockstep with business, we communicate better, appreciate each other's vision and mission, coordinate programs in a more effective and efficient manner, and have better success. The point is, whatever change is being implemented in the way the business is managed and operated needs to be duplicated in the safety and health function. It cannot remain the way it always has been done. That won't work!

Let's look at the credit side of the ledger. What are the advantages to business if we move away from the way worker safety and health has always been done? A Japanese manager on a tour of an American production facility told me a saying by Taiichi Ohno, the father of the Toyota Process (also known as Just-in-Time manufacturing): "You can't make quality in a pig sty." Let me interpret that remark and expand it further: No aspect of business stands alone. Everything is dependent on everything else. For example, you cannot have high quality without efficient production. You cannot have excellent cost control without effective management. You cannot have high morale without high participation and active communication. You cannot have exemplary housekeeping without worker dedication to the organization's vision and mission. You cannot have efficient utilization of your people resources without dynamic strategic planning. You cannot have excellent worker safety and health without all of the above. Too many times, we lose sight of the interrelated nature of all the critical aspects of our business. When a wheel comes off, we try to solve that particular problem without addressing the other interrelated aspects that go hand in hand with it.

Too many times we consciously devalue some critical aspect of our business and refuse to see that by doing so, we create an imbalance which is unmanageable. The simple fact of management is that if we realize the total interrelationship, set equal values, and manage all aspects in the same manner, management becomes easier, the business runs smoother, has fewer upsets and is more profitable. This isn't Management 351, it's a management reality check. If any manager thinks that he or she can run any business and ignore or devalue any critical aspect of that business, he or she will not be successful—nor will that business. Make no mistake about it. Worker safety and health *is* a critical aspect to any business.

BUSINESS PRIORITIES

When we discuss critical aspects of a business, we are really talking about business priorities. Let's be perfectly clear about something. In any business, who sets the priorities for that business? Management does, the line organization. Have you ever studied mission statements or non-financial objectives of businesses? I have, a lot of them. It's one of the first things I ask to see when I visit a facility or consult with any business. One thing that has always stood out between American and Japanese mission statements and non-financial objectives is the issue of worker safety and health. Worker safety and health is *always* included in Japanese business documents. It is seldom, if at all, included in the same documents of American businesses. The message that seems to be there is that worker safety and health is a priority in Japanese business. There it is seen as a critical aspect, a high priority of business. In America, it would appear that worker safety and health is a side-note or an afterthought, not a priority at all. If management develops a mission statement and business objectives to guide the business but doesn't include safety and health, why does American business wonder why no one outside of management thinks that worker safety and health is important to management? It would appear that American management talks a good line, but they seldom "walk the talk" when it comes to the safety and health of their workers. Look at the evidence. Even the documents which guide American businesses, their own mission statements and non-financial objectives, bear witness to that truth by the omission of worker safety and health issues.

A Fortune 50 company recently began an impressive and ambitious effort. Their corporate environmental, safety and health office began to write guidance manuals for their numerous facilities. It was a great idea. Through the manuals, a large step could be made toward standardization of programs. Because of the complexity of these manuals, a priority list was developed. Work would begin with the first manual on the list and when through, printed and distributed, work would begin on the second, etc. until all manuals on the list were complete. The plan called for development of four manuals. What was the subject of the first manual? It covered Environmental Management. How about the second? It included Emergency Response Planning Guidelines. The third? That one addressed guidelines for handling spills and releases. When did they get around to worker safety and health? Oh, that was manual number four. What were the priorities of this Fortune 50 corporation?

Management has long "talked a good talk" concerning worker safety and health, but their "walk" never seems to meet the expectations from the "talk." After all, was it workers who invented the phrase, "Safety is number one?" No, it was management! The number one objective of any business in America has to be profit. In our free enterprise system, there is no other way it can be, at least not for very long. So, in American business, profit is number one. That point is not at issue here. What *is* at issue are the other business priorities and the interrelationship of those critical aspects of running a business. Interrelated here means to be in *balance*! The critical aspects of conducting a business have to be in balance if that business is to be successful at their number one objective, making a profit. Sure, it's a long-haul concept. But, if any aspect is devalued or ignored for any extended length of time, the imbalance will throw *all* aspects into turmoil. It's a business fact. It's a management fact. And if worker safety and health is hidden, not included in some of the most important business guiding documents and in their objectives, the imbalance is obvious. Why then, isn't it obvious to traditional American management?

PRIORITY TO VALUE

Let me add to my point about priorities. There is also a negative aspect of assigning priorities. Priorities tend to change depending on what's "in our face." For example, you may have several priorities in your life. One may be to send your kids to college, so you begin saving for that eventuality. Another priority may be to put food on the table every day. What would happen if you suddenly became unemployed? Let's say you couldn't find another job for some time. How would this impact your priorities? It would cause a reassessment, wouldn't it? The college fund would become less important than the necessity of putting food on the table. After all, it is easy to argue that your children wouldn't need the college money if they starved before they got there.

In response to having our economic pants kicked, America is changing the way we think about quality. Sure, we always thought it was important. It was one of our priorities in business. However, when our quality remained constant and the quality of products made in other countries became better and better, we had to reevaluate the way we viewed quality. It was a classical example of our daily reassigning priorities. We realized, however, that the "ship it anyway" compromise on product quality could not continue. In other words, we could not allow the status of quality to be reassigned by what was in our face at any particular time. Quality had to be raised to a higher level. It had to become a value.

The difference between a value and a priority is rather simple but important. Priorities are continually reassessed depending on circumstances. Values are non-negotiable. They are never reassessed, changed or compromised. Look at getting up in the morning as an example. Upon arising, we have several priorities while we get ready for work. One priority may be to shave or to put on makeup. Another may be to eat breakfast. Another may be to read the newspaper. What happens when we arise fifteen minutes late? The priorities are reassessed. The newspaper reading gets put off until we return home after work. What happens when we wake thirty minutes late? Priorities again are reassessed. This time, we shave or put on our makeup in the car on the way to work. When circumstances change, priorities are reassessed.

Getting dressed in the morning is not a priority—it is a value. If we arise forty-five minutes late, do we choose to get dressed when we get to work? No, getting dressed on the bus or in the elevator at work is non-negotiable. It is a societal value. We always get dressed before we leave home—even if it makes us late for work.

Making worker safety and health a value is an important part of creating a value-adding environment. It must become non-negotiable. Otherwise, safety as a priority is placed in an unavoidable position of being compromised or perceived as being compromised. Real or perceived, the impact on creating a value-adding environment for worker safety and health will be catastrophic to any efforts. Safety and health must become a value to the business.

THE NEW AMERICAN BUSINESS

The era of single-issue businesses is dead. No longer can we view economics as the only aspect that will make us profitable or not. Businesses and their workplaces are more complex today than ever. Change is rampant. We must change if we are to be successful or survive. Change is becoming more and more rapid. We no longer have the luxury to sit back and change slowly, if we choose to. We must become change masters and change quickly by intuition alone. Quality is no longer a choice or priority. It must be a value if we are to compete in this global economy. As Joel Barker put it, "If you don't get it (quality), you may not survive!"

Other aspects of the workplace can no longer be viewed as unimportant or of secondary need. Everything in an efficient, rapidly changing, highly participatory workplace must work in carefully orchestrated precision. Everything is interdependent. You cannot fail in one and succeed in another. Worker safety and health is one of those critical business aspects. Make no mistake about it, we cannot hope to reconquer our position in the global economy without excellence in worker safety and health. And just as the system for worker safety and health had its roots in and in response to the workplace, the rebirth must also begin here. To win at worker safety and health, the war must begin in our workplaces with critical changes in the way we manage and think. Without this fundamental seeding, we cannot begin the rest of our efforts. It is that important.

SUMMARY

Current system problems—what we should stop doing:

- Defining safety and health as a totally economically-based program for cost-containment.
- Placing the responsibility for safety and health in a staff-oriented function without direct authority over day-to-day issues.
- Dealing with safety and health coverage ratios and corrective measures in an economically justifiable way exclusively.
- Perpetuating the "Safety Is Easy" paradigm.
- Dealing with worker safety and health as a third-rate priority.

Fixing the system—what we should be doing:

- Placing accountable responsibility for worker safety and health into the line structure.
- Making the staff safety and health a service-related function to the line based on need and associated hazards of the particular industry or business.
- Make safety and health a non-negotiable value in the organization's culture.
- Recognize that safety and health responsibility starts at the top of the organization and carries down through all employees.
- Making safety and health a team-based concept with high participation at all levels.

3

SAFETY AND ORGANIZED LABOR

I had a good friend while working in a plant with a unionized workforce years ago who was the union's safety committee chairman. I was in safety and health management. We got along well because we shared a common goal—improving worker safety and health in the plant. We both sat on the Joint Union and Management Safety Committee (JUM). However, 95 percent of the progress we made was done "unofficially" by the two of us. In his words, we didn't exactly follow the rules that were created by the contract. But, in the two years we worked together "unofficially," we made many significant advancements for safety and health at that plant. We were both frustrated with our official roles and the slow progress of the JUM. There were union members on the committee who were obviously there for their own political interests and management members who viewed this committee as only a confrontational aspect of the grievance process. Interestingly, when my union friend and I attended formal committee meetings, we behaved differently, each wearing his "official face." No wonder we worked together on the outside to get things done. The safety and health of the workforce was just too important to both of us. And the JUM was much too political and ineffective to allow us to get things done.

After that, he was elected to the local union's presidency. Three things changed drastically. He never was without his union "face," the momentum of advancing safety and health in the plant slowed to a snail's pace because

> the JUM became the focal point, and we became distant acquaintances. Being on different sides of the fence became *the* issue that eroded our friendship and our ability to effect change.

This is not a discussion of whether or not you like or agree with labor unions. It is also not a discussion on the pros or cons of becoming unionized or joining a union. Those are both personal positions and decisions. Like OSHA, unions exist. But you would have to be from Mars not to be understanding and empathetic with the reasons unions began.

ORIGINS OF ORGANIZED LABOR

In the Beginning. . .

Does the name Samuel Gompers sound familiar? Most junior high (or middle school) students recognize his name from their study of American history. He is generally accepted as the father of organized labor in the United States. Some will passionately argue that he wasn't and that someone else was. That's not the point. For the purpose of our discussion, it might as well have been Elias Birdsong or Elaine Kilpatrick. It just doesn't matter because, in reality, none of them was the "father" (or "mother," if you wish) of organized labor. Then, who was? It wasn't as much a *who* as a *what*. The true "father" of organized labor was poor business management. Gompers, or any other organizer, merely responded in a logical manner to an irrepressible force. The point is, if it hadn't been Samuel Gompers, it would have been someone else. Poor management was such an impressive reason that whoever was the first to organize workers is really unimportant. What does matter is what caused unions to get started in the first place.

Valuing Products Over People

Remember our earlier discussions about the history of worker safety and health in America? Did any one particular theme strike home concerning how business and management in the early 1900s thought of workers? Would you say that it was with respect and high value? Was management concerned with how the workers felt and if they were happy productive workers? Were they concerned with them as human beings? You'd have to stretch the definition of human being pretty low on the animal scale to be able to say yes. In fact, management at that time probably felt more compassionate about their family pets than they did about their workers. It's hard to argue anything else than that management in the early 1900s considered workers merely a "means to an end." The "end," of course, was profit. It took a certain number of workers to make a certain amount of product. If production was to increase, the number of workers would also have to be increased. If the number of workers decreased for some reason (including death or serious injury), more workers would have to be hired so that the needed amount of product could be made. This was one of the ugliest points in our nation's business history.

Unfortunately, management's traditional thinking hasn't changed much. Peter Drucker states in his classic book *The Practice of Management* that economics is the number one job of management. Business economics has two possible outcomes at the end of any fiscal period—profit or loss. The important point is that management is still focused on economics—the next quarter's balance sheet. What did Drucker suggest should be the most important consideration for choosing staff personnel (in his book *The Frontiers of Management*)? Operational experience. Why? Because operational experience guarantees business knowledge—that is, knowledge of what makes the business profitable (economics). It's a cost versus value concept. It isn't rare. It isn't hidden. It is in wide practice in American business today.

What were the results of this type of management? What possible reasons could workers have had to unite? Other than that the workplaces were unsafe and workers had to put up with extremely poor working conditions, low wages, long hours and no job security, I can't think of any.

It's no small wonder that workers banded together in an effort to make their collective voice heard. In retrospect, it's a real shame it had to happen. It's a shame because it was a black mark in our nation's proud history. It's also a shame because management's actions of that time were totally unnecessary. The carnage that occurred due to attitudes and lack of action concerning worker safety and health was abominable. This was based, of course, on total contempt for workers in general. One cannot look back and not become sickened by that period in our history. Today business as a whole wonders why workers and the public remain suspicious of their motives. One bad experience really does wipe out 1,000 good ones, and people have extremely long memories when it comes to negative things.

IMPACT OF ORGANIZED LABOR

What impact has organized labor had on worker safety and health? It has been both positive and negative. It has been positive from the aspect of being a major player in the worker safety and health movement. One simply cannot deny the positive influence of organized labor not only in bringing about worker safety and health emphasis in business, but also in helping mold and shape the form of today's safety and health work environment. In many instances, organized labor forced businesses to implement worker safety and health programs where there were none. They pushed for a national approach to mandating worker safety and health. They helped draft safety and health standards and regulations. They "stood up and were counted" for worker safety and health during many tough times over the past hundred years. On a facility scale, many hazards in organized shops have been identified and corrected through union actions. These are all positive.

But there are negative impacts too. These are much more difficult for us to talk about. They tend to be emotional issues. But in fairness, we need to look at both sides of the issue. What are the negative impacts of organized labor on worker safety and health today? They fall into six categories: 1) unions de-individualize worker safety and health, 2) they tend to politicize it, 3) contracts commonly get in the way of worker participation, 4) unions tend to slow improvements via politics, 5) unions devalue individual needs and abilities for the collective good, and 6) by their paradigms, they reinforce the polarization between labor and management (and the status quo). Each is a significant impactor on worker safety and health.

De-Individualizing Worker Safety and Health

First, unions de-individualize worker safety and health. Unions have structure, just like a business does. They have to or communication up and down the union hierarchy would be impossible. But this structure places a "middle man" between worker safety and health issues. Arguably, it allows for better and more direct communication on safety and health issues or establishes a communication link when individuals don't want to come forward. However, a worker with a particular safety or health concern generally has to tell someone else who, in turn, has to tell someone else or who negotiates the "settlement" of the worker's issue. The problems with this approach are many. It can take longer to resolve the issues. The identification and urgency are greatly dependent upon the worker's "clout" in the union structure and/or communication skills. Resolution is often incomplete or forced into a back-and-forth cycle before it is resolved. Concerns can be overlooked because they don't "measure-up" to other current safety and health issues or other high priority union issues. And, because of the removed resolution process and the time it may take, it places a negative barrier between worker and management. In the union structure, the "middle man" position is unavoidable, but in application to personal issues such as safety and health concerns, it is very often cumbersome and inefficient.

A union worker in a steel-making facility felt that a particular job he was asked to do was unsafe. It was a legitimate concern. As per protocol, he complained to his department union steward. The union steward and the department union safety representative went to the department superintendent. The department manager listened to their complaint and afterwards called in the concerned worker's foreman. The manager wanted to collect as much information as possible in as short a time as possible because work had come to a stop. The facility safety supervisor and safety engineer for the department got involved. One hour passed—two hours. Finally, they called the union steward to see if their solution would meet the union's concerns. It did not. With things getting ugly at the worksite, the job was postponed until the next day. The union steward, the union safety representative, the facility safety manager and the department safety engineer went to the worksite to try to resolve the issue. After about two hours, they reached a solution. The necessary safety equipment was acquired via one-day express delivery. The safety engineer and the foreman gave the safety equipment to the concerned worker. It didn't address his original concern. After all this effort, no progress had been made.

How would this issue have been resolved in a non-unionized facility? The concerned employee would have shared his concern with his direct supervisor. The problem would have been resolved at the worksite between them or with the help of safety people. The whole operation would have taken less than an hour and it would have addressed the worker's concern. It follows the KISS principle (Keep it Simple Stupid). In a unionized shop, however, it simply can't and doesn't work that way.

Politicizing Worker Safety and Health

Second, unions tend to politicize worker safety and health. Let's face it. Politics has become a way of life in almost everything we do. If you want me to do something for you, you need to do something for me. It's a "Quid Pro Quo" concept. It's how our government works, especially the legislative part. How well (or efficiently) does Congress work? No argument. It works very inefficiently and, at best, very poorly. One of the major reasons is because of the politics that are as much a part of Congress' culture as the "power structure" is. It's one of the major arguments for term limitation. Limiting terms, it is thought, will limit the amount of politics, thereby, make Congress more efficient and over time change the culture of Congress. Just as politics makes Congress inefficient, it also makes resolution of worker safety and health issues inefficient in unionized workplaces. When any culture is set up so that it "trades" something you want for something else, it *cannot* be efficient. And, in today's business world, efficiency equals survival. Unions cannot expect to be immune from this changing picture.

Whether you look at the events surrounding contract negotiations, the grievance process or the JUM, politics is unavoidable. In each instance, the union has their views and management has theirs. One side seeks to gain credibility and agreement on their issues and discard their opponent's issues. It's called negotiation. First, unless you have all the power or "cards," it's not a "Burger King." In other words, you can't have it your own way. You get some and the other side gets some. How much is totally dependent on your power position, the weakness or strength of their issues, the balance or imbalance of negotiating skills between parties, and if anything is "owed" from past negotiations or issues. Second, the only way one side makes progress is to give the other side something they want and feel is significant. If neither side is willing to give something to get something, a stalemate results. In historical terms, it's called a strike. Third, more times than not, what you get in the end is often not what you wanted in the first place. It's one of the unwritten rules of good negotiations. Even if you don't get what you want, you get a partial win if the other side doesn't get what they want either. Fourth, when negotiating for others, you never get a "great job" from the team. Often you're lucky if they ratify the agreement. And fifth, little issues get lost in the fury of negotiating big issues. Too often, significant safety issues or local concerns get ignored in the belief that if we gain something close to what we want on the bigger issues (job security, pension, pay, etc.), others can deal with the little ones on a local basis if they still think they are important. This is no way to deal with safety and health issues if we are to make progress in worker safety and health. Safety and health issues are as important to workers as pay or benefits.

Safety and health concerns must be met with sincerity. It's a human, caring issue. Politics is about bargaining—an inefficient squabbling behavior. Sincerity and politics are antithetical. What is your response when you hear a politician? Suspicion, isn't it? Sincerity and politics do not mix. It's one or the other, not both. When your child comes to you with a problem—important to him or her—do you play politics with it? Of course not. It would be demeaning to the child and devalue him or her concern. How is it then that we can play politics with safety and health issues? Unions were founded on the need for worker (human) rights. This is a foundation based on sincere concern for humans and the human spirit. Reducing safety and health issues to politics is demeaning to the foundation of the labor movement. We simply can't allow it.

Contract Limiting Participation

Third, the union contract itself gets in the way of worker participation in safety and health. Our business world is changing. In order to be more responsive to business demands and changes in technology, business *must* achieve high participation by its workers. The old philosophy of "management thinks" and "workers work" no longer holds. Contracts change slowly, very slowly. This happens because of the politics that are involved and because unions don't want to lose anything that they have gained over many years. You can't blame them for that. But, if America's businesses are to be successful in this rapidly changing global economy, greater worker participation must happen. This removes some of the built-in worker "insulation" and this is frightening to a lot of unions and workers. Worker participation in safety and health goes hand in hand with this change in American business. It has to happen.

Labor unions in some other nations think differently. Their thinking often transcends the American "defined by contract" and it's a "us against them" mentality. A safety professional was visiting Japan in an effort to learn what they do differently in order to have such exceptional levels of worker safety and health. While visiting a large manufacturing facility, he asked senior management if he could speak to the union president. Enthusiastically, they agreed that this would be meaningful to the American's efforts. A meeting was arranged by management for later that day at the corporate offices. When the American showed up for this meeting, he was met by senior management who escorted him to a lavish conference room where the union president awaited. Drinks and hors d'oeuvres were served. He noted a genuine openness and respect between the senior management and the union president. Left alone, the American and the union president began talking. Many questions were asked and forthrightly answered. One answer, however, was strikingly different from the American way of thinking. He asked, "How do you and management resolve pay and benefit issues?" The answer was simple but deafening. "If we want a bigger piece of the pie, we must help management get a bigger pie."

In America, this story would have been much different. The senior management would not have been enthusiastic about a visitor meeting with the union president. They certainly would not have arranged it. The meeting would take place at the union hall, not at the corporate offices. No lavishness would surround the meeting. No exchange would take place between senior management and the union president because they wouldn't be meeting with the visitor together. If they met together, there would not be the same openness and respect between parties. And very importantly, the answer to the visitor's question would have been very different.

Slowing Improvements

Fourth, unions tend to slow improvements. The "middle man" structure and the unavoidable politics that are part of the union structure and culture simply slow things down, including the resolution or correction of worker safety and health issues. American business must become quicker and quicker if it is to succeed in the global economy.

Recently, members of an engineering group from a fast-paced facility attended a large and traditional American corporate training session. One of the presenters talked about workplace changes and the importance of safety and health input in resolving safety issues prior to construction. He outlined a lengthy sequence of activities, meetings, reviews, sign-offs, and walk-throughs that his facility used to assure that all safety and health issues were resolved before the power was turned on. One of the fast-paced facility's engineers asked, "How long does that take?" "Usually between 14 and 18 months," he was told. "How can you get anything done? How can you keep up with the global competition? Things are faster today than ever. We have to have large changes implemented at our facility in three months, six months, tops." The engineer continued. "At our facility, safety and health staff walk with us through every aspect of the design and implementation. They coauthor the project. It's the only way we can make changes as quickly as we do. It is also the only way we have been as successful as we have." The room was very quiet. They didn't understand what they just heard. It, of course, was a message from the future of American business.

Today's and tomorrow's world will not respect slow-paced change. Just as technology is changing before our eyes, changes in the workplace must happen at the same rate if America is to regain a leading position in the global economy.

Devaluing Individual Needs and Abilities

Fifth, unions devalue individual needs and abilities. This has been the unavoidable result of the "all for one (union) and one for all" premise of organized labor. It isn't supposed to be about individuals. The collective good is what it's all about. That isn't necessarily bad. Without this collective "clout," worker safety and health would be worse off than it is today. But today's business wealth, the resource that will insure our economic success in this global environment, lies in the individual gifts and the associated needs of workers. It does *not* lie in management. Sure, management must lead the way into the future, but the most important contribution that management can make to insure America's business future depends on creating an environment where individual worker participation is free and open so that these gifts can make us successful. But creation of this work environment is a two-way street. Organized labor must *also* change in order that this free and openly participatory work environment can occur.

A union pipefitter had a novel idea while working on a project. It was one of those things that kept going through his head. Mentally, he could see the idea in action. He could see the improvement. He also knew that he couldn't implement it because it was electrical. He was a pipefitter, not an electrician. In the union structure (contract), he was helpless to implement it. So he tried to work with an electrician to get the idea implemented. For some reason, the electrician didn't see the same value in his idea. Maybe it was because the idea came from a pipefitter, not an electrician. Maybe it was jealousy. Maybe he just didn't understand. In any event, he wouldn't even entertain the thought. Frustrated, the pipefitter went home and built it in the shop in his garage. Not only was it a good idea that worked, it was patentable. Getting the patent took him a year. He spent the next year looking for a manufacturing company that liked his idea. When he presented his idea to a small electronics manufacturing company, they saw a gold mine. They began to manufacture it. Two years later, the company the pipefitter worked for was able to purchase the idea and use it. Sure, the pipefitter made out okay. He makes money on every unit the now mid-sized electronic manufacturing company sells. But what would have happened if this idea meant the difference between success and failure to the pipefitter's company? Having to wait more than four years to be able to use it and gain the improvements is ridiculous. Rapid change is the key to tomorrow.

Reinforcing Polarization Between Labor and Management

Sixth, unions reinforce polarization between labor and management, the status quo in the workplace. Let's have a reality check here. As the paradigms currently exist within organized labor, it is in a union's best interest to continue this polarization. It is, in fact, a large source of union power. It forces things to go through union channels. It reinforces an "us versus them" environment. It reinforces the "we (the union) are the only ones that care about you (the worker)" paradigm. And, looking at management practices over the past hundred years, that paradigm has been well supported by management actions as well. And, reinforcing the status quo, the "business as usual," or "that's the way it's always been," keeps American workplaces from moving into the next century, including in the area of worker safety and health. Obviously, something has to change.

Let's return to the example of the traveling American safety professional in Japan. Senior management was enthusiastic about the visitor meeting the union president. In fact, they arranged and hosted the meeting. The union president was warm and very open. There wasn't a single question the visitor asked that was refused or answered in a "skirting the issue" manner. These are eloquent statements of four things that are very different in Japanese business compared to those in America. First, there is an openness between management and labor. Second, there is a lot of mutual trust. Third, there exists a partnership between labor and management, each understanding its own role and respecting the other's. And fourth, there is true participation by workers in making the business successful. Sad to say, these things are not common in American workplaces, especially where workers are organized. In most, they are nonexistent!

After all, our system was built on labor and management not getting along. Sadly, rather than trying to fix it, we have expanded the problem by placing a third party, a part of our inefficient and political government, in the middle. I'm talking about the National Labor Relations Board, the NLRB. The NLRB was created because labor and management couldn't get along. It was created to level the playing field and oversee management and labor—in other words, to make sure that neither side was gaining an advantage. Like a referee at a professional wrestling match. The problems this extra layer has created include more regulations, the time it takes to mediate problems, and like most federal agencies, extreme politics. One time the NLRB will vote one way, the next another way, very much like the professional wrestling referee. Think of the savings to business and to the taxpayers if management and labor could get along and be on the same team for American business and worker safety and health.

As an example of how the NLRB has become an impediment to improving the system for worker safety and health, consider the *Electromation* decision in 1992. The question that came before the NLRB concerned employee involvement in company committees, such as safety committees, that were not explicitly agreed to in labor contracts. The Board found that "employer involvement" teams were "labor organizations" under the National Labor Relations Act, specifically, Section 8(a)(2) of the Act. The Board concluded that Electromation had violated the "terms and conditions of employment" wording of the Act.

It's a simple example of a governmental action (the National Labor Relations Act and the creation of the NLRB), becoming part of the problem. Born totally out of management and labor's inability to get along, the system that directly impacts furthering worker safety and health becomes additionally encumbered, inefficient, and ineffective.

ROADBLOCKS TO CHANGE

In order for positive change in worker safety and health to happen, there are some significant roadblocks we have created that must be dismantled. We must focus on two areas—management-labor relations, and the resulting bureaucracy that has sought to "balance" this workplace relationship. To say which one needs to be worked on first is a "chicken or egg" argument. Personally, I feel that you need to solve the management-labor problems first before you can attack the bureaucracy. After all, it is reasonable to conclude that the bureaucracy was created as a direct result of poor management-labor relations in the past (and continuing today). Without solving the problems in the workplace first, any change in the bureaucracy would be impossible. Others, however, argue the reverse. If you don't change the bureaucracy first, the existing protectionistic laws and the Board will derail any positive efforts between management and labor that fall outside the edges of what the NLRB might call the "field of battle." They've got a good point. Perhaps then, we should work on both at the same time.

Dismantling the Protectionistic Bureaucracy

Obviously, if we are to chart new waters for management-labor relations and move from combative and suspicious attitudes to cooperative and participatory ones, we must rewrite the labor laws. American business must change to compete in the global economy if we, as a nation, are going to have a viable economy in the future. It is unrealistic to expect old, unchanging, protectionistic labor laws to do anything but provide significant roadblocks to our ability to compete by forcing the status-quo down the throats of business, management, and labor as they seek change. This effort can only begin with questioning the validity of each law's purpose, measured against today's realities. Where the laws' purposes are no longer valid, we must have the courage to throw them away. Where the purposes are still valid but aspects of the laws do not interface with today's or tomorrow's needs, we must change them so that they do. In a changing business environment, one in which our nation's economy is at stake, we must demand a dynamic environment for labor law. Dynamic change cannot coexist with static environments. This conflict will only choke the life from advancement and seal our fate as a second-rate economic power.

Who must take on this effort? We must do it together—management, labor and government. It is for this reason that I feel strongly that a new relationship must first be built between management and labor before a new cooperative effort with government can have any chance of success. It would be too challenging to build positive relationships on two fronts. Only through leading the way by first establishing a new relationship between management and labor, can we change the relationship with the bureaucracy.

Creating a New Management-Labor Relationship

In theory, creating a new management-labor relationship is easy. In practice, however, it will be anything but easy. Many changes must occur. A long history of poor relationships must be prepared. Traditional beliefs and practices must be discarded. Success in this effort revolves around two key words—trust and respect. If we can build trust and respect between management and labor, two-thirds of the battle is won.

What gets in the way of building this kind of open, trusting, respectful, and participatory relationship between management and organized labor in America? Part of the problem lies in organized labor, and part in management. As we've discussed, there are serious impediments within the union structure and culture which throw roadblocks in the way of such a relationship. Some of this is a status quo or "not giving up what has been fought so hard for and won." Some of this is a protectionistic issue. It's pure fear. In other words, if we (the union) change, can we survive?

The Trust Issue

But one of the largest roadblocks is a trust issue. With so much history, I truly doubt if organized labor trusts management to be different, because in such an open relationship, management would definitely *have* to be different. This is not politics. It is not an "if I change then you have to change too" political issue. It's a fact. To build such a relationship, both labor and management would have to change.

Recently, on a business trip, I arrived in a major American city very late. My schedule had necessitated that my flight be moved to the latest one that day. The flight was full and late leaving. I was tired. It had been a long and exhausting day. After renting a car, I drove the 25 miles to the hotel where my secretary had set up my reservations for the night. Parking the car, I walked around the hotel to enter at the front. There was a picket line. The workers were on strike. Why were they on strike? It really wasn't important to me. Turning on my heels, I returned to my car. I found another hotel for the night.

Why did I do that? Certainly I knew that I was not in any danger crossing the picket line. The striking workers were very polite. Am I a "union lover?" No, but I don't hate them either. What was the problem, why did I refuse to cross the picket line? The answer is simple. I knew immediately what the problem was and I refused to collude. It didn't matter what the workers were striking for or against. It could have been wages, benefits, safer working conditions, or that they weren't getting enough hotel mints. I didn't care because *that* wasn't the problem. The problem was communication.

Improving Communication

I have a simple philosophy concerning communication in business. If it is management's responsibility to lead the business, it is their responsibility then to maintain communications with the troops. After all, what kind of army would you have if the generals never communicated with the foot soldiers? Communication is a management responsibility, *not* a labor responsibility. What happened to precipitate this strike was that communication broke down, wasn't effective, or wasn't initiated. And whose responsibility was that? Right, it was management's responsibility.

How did I know that it was a communication problem? If communication had been successful, there wouldn't have been a strike, would there? But, studying American management, this failure to communicate is very common. In reality, it's a throwback to the "way we've always managed." American management has long been expert at what I call "what" management. In other words, management tells the workers "what" they want them to do— "I tell, you do." It's a lot like the way we talk "to" our children. "Clean up your room." "Take out the garbage." "Go to bed." "Be in by 11:00." It comes from the long-term, deeply entrenched management philosophy in America where it is believed that management "thinks" and labor "does." And because our businesses have been successful using this "what" management for so many years, it has been nurtured, protected, and still is believed to be the best way.

I don't know how many times I've been told by management, "You can't tell them (workers) that!" "Why not?" I always ask. I've never received a good answer to that question. It's probably because they don't know either. It's just the way it's always been.

Importance of "Why" Management

What changes in management communication are necessary to form a closer relationship with workers? It's called "why" management. In other words, just don't tell workers "what" you want them to do, begin by telling them "why" it needs to be done. Human beings are programmed with a basic need to know why we are doing something before we can do it. This doesn't apply just to the workplace. It is a basic need in all aspects of our lives. The why can be as simple as "Dad said so," "the law requires it," or "the boss said to do it." But it can be very complex also. The reason why can be obvious and easily seen. It can also be hidden or based on pure speculation. The point is, why we do something is a critical piece of information before we can act.

Telling someone "why" rather than "what" also tells them how important they are to us. We always insist on telling the important people why. Why an investment is solid. Why the boss should back our idea. Why our community leaders should support or reject a project. We insist on telling important people why. However, we tell unimportant people only what. Think about this. It's true. Do your kids think they are important or unimportant by the way you talk to them? "Finish your dinner." "Take out the dog." Does this also fit with the way management talks to workers? Remember the picket line at the hotel I was going to stay at? Answer this question about management communication yourself.

What has changed in American business that has made "what" management no longer good enough and "why" management necessary? Has the American workplace changed in the past twenty years? Would you say that it is a whole lot more complex? Have American workers changed? Would you say that they are a lot more knowledgeable and aware today? Let's not get into an argument whether or not "what" management has *ever* been successful. But you've got to admit that the workplaces and workers of today are vastly different from those of years past. Back then, "having a job" or "because the boss said so" were reasons enough to act. Today, they aren't! And because of that, management must change to "why" management if it is to be successful in this new workplace. Period.

That statement runs headlong into some longstanding management paradigms concerning how much workers should know or need to know and how open and cooperative management must be in order to maintain discipline, especially if those workers are unionized. These paradigms are no more than myths, ghosts from management past. Does it make any sense, knowing how important worker participation and input is to the future of any business, to not communicate fully with them? There are a lot of deeply entrenched paradigms. "They wouldn't understand everything." "That's upper level information only." "Workers just don't need to know everything."

Valuing Worker Communication

There are also some risks to management if "why" management is used. For example, if workers know more, they may not agree with the leadership or conclusions that management provides or makes. But, let's face it. Hearing two sides of an issue always provides a much clearer picture and usually results in better decisions. That's why both sides of an issue are included in our judicial system. Also management sometimes gets myopic, only seeing things from their perspective. Wouldn't it be nice and of great value to management if they could truly see things from a workers' perspective, not just from their own viewpoint or what they think the workers' is?

I happen to be an optimist. And my optimism usually serves me well. I believe that workers aren't stupid. Given enough information and fully understanding the ramifications, they can help make excellent decisions, or accept difficult ones that management has to make, especially when those decisions affect them. Why else do you think problem-solving teams that have workers in them are so effective? How else can you expect to stop this unnecessary waste of American productivity called "strikes?" Sure, the workers may have a different opinion. They may even have a different vision based on their needs. But that's perspective. That's critical input if management is to lead into the future. And using "why" management, respecting the intelligence of and input from workers, is the only way to get there. Sure, "why" management is more work—at least in the beginning. But it avoids so many problems and ill feelings that there really isn't any other way to manage.

"What" management maintains the polarization and protects the status quo. It's degrading to workers. It is the direct cause of management/worker conflict, including strikes or work stoppages. It is antiquated and should be buried along with the way America used to run its businesses. As with business, leadership in change lies in the court of management. Management must change. It must lead the way to tearing down the historical wall between management and labor. A good beginning is coming to grips with the reality that poor management of the past is really the source of most of management's problems in business *and* with labor. Management created the monster that exists today. Trying to blame labor, government, the global economy, or any other outside issue only delays solving the problem.

IT TAKES TWO TO TANGO

But, it isn't just a change that has to happen in management. Sure, management has to take the lead and change first. After all, they not only started the problem but it is the responsibility of management to lead. But labor, especially the unions, must also change. They must recognize the importance of their role in this process. Traditional roadblocks to honest communication and politics must be discarded for the betterment of American business. This result, of course, accomplishes a major piece of union business—making jobs more secure and not having them lost to a foreign country. Management must lead in this change effort but labor must also change. One side changing without the other only dooms the entire effort.

The problem is, of course, one of trust. Trust between management and labor must be rebuilt, but this can only be accomplished one step at a time. The unions must realize that the accepted and nurtured separation between management and themselves will not work in tomorrow's workplace. Unions must also realize that worker participation is the key to our business success. Barriers to that participation must be removed and more direct communication pathways between labor and management opened.

In today's and tomorrow's successful businesses, there must be a mutual trust between labor and management, and workers must be fully supportive, involved, and active in the business. But this cannot happen just because we think it's a good idea. We have to recognize and accept that it doesn't exist and that to be successful in the future, especially for worker safety and health, it must be created. Both management and labor must be committed to changing the ways we have always done things so that we can create this workplace. We must realize that to change is to risk, but that not risking will not advance safety and health and will also doom our American business world. Then, we must see the journey through, fine-tuning as we go, but going together.

SUMMARY

Current problems—what we should stop doing:

- Perpetuating the antagonistic relationship between labor and management.
- De-individualizing individual needs and aptitudes.
- Being a barrier to participation.
- Politicizing worker safety and health.
- Perpetuating labor laws that decrease our chances of success tomorrow.

Fixing the system—what we should be doing:

- Following management's lead, opening new avenues of candid and sincere communication.
- Recognizing that success of American business tomorrow means a commitment to team and individual participation today.
- Changing the laws that limit participation and the NLRB, if necessary.
- Management communicating "why" not just "what."
- Streamlining union hierarchy to provide quicker and more effective resolution of issues and concerns.

4

THE COMPENSATION QUAGMIRE

> A new employee was employed at a facility for five days; of those, two were missed. The employee called in sick. That was more than four years ago. They company is still paying her compensation. What happened? The worker claimed to be sick on the second and third workdays and missed two of the three days that had been set aside for new employee training. The fourth day was spent trying to catch the worker up with the new employee group that she was hired with. On the fifth day, when they began acclimating the new employees slowly into the workplace, this new employee bent over, felt a pain and went home. She told no one. The employee just disappeared. The next Monday the employee called in to her supervisor and told him that following the advice of her doctor, she would be off for a while because of back strain. That employee has still not returned to work.

If you talk with *any* employer, they will tell you stories like this one. These cases are not uncommon. Is this prima facie evidence that America's workplaces are unsafe? No, don't get lost in the trees like too many others have. In reality, the individual cases are unimportant. The injuries and supposed injuries are also unimportant to this discussion. What is important here is the forest, the part of the system: the workers' compensation program. Its problems are very convoluted and very

67

complex. It has many problems which result, time and time again, in system failures.

An employee came into the medical department of a facility a few years ago. He needed help. He had supposedly been injured four weeks earlier. Wanting to be a "good" employee and not perceived by his work crew as a complainer, he didn't tell anyone. He didn't tell his supervisor. He thought the injury would go away if he treated it. Well, it didn't. By the time he finally sought medical help at a local emergency room, the injury was severe. The problem that he had brought to the medical department was twofold. First, the attending physician was insisting that he take time off work to heal. As with most workers' compensations systems, that would mean loss of income for the first few days off work. But the system had failed and that caused the real problem, his second problem. He had put "yes" on the hospital's treatment questionnaire where it asked, "Was this a work-related injury?" Trying to keep the injury "quiet," he had filed a claim against his accident and medical insurance. Now they denied coverage because it was work-related. The state workers' compensation office had also denied coverage because they had not received the necessary forms from his employer. So there he was, stuck with bills that neither workers' compensation nor his insurance would pay and being told to take time off from work. That, of course, would mean no income at all. He had a considerable problem. The system had failed again!

THE CALL FOR NEW THINKING

You might say that our society has "opportunists" and "idiots" in it. There is no way that any program can be controlled so tightly that some "opportunist" will not be able to find a loophole. The worker who violated the workplace rules by failing to tell management that he had been injured only got what he had coming to him.

This is "old thinking." Unfortunately, we've been using "old thinking" for years. That's why we haven't been able to solve the problems with the workers' compensation system. In order to make the problems visible and solutions possible, we must think differently. To use a phrase used earlier in this chapter, we must focus on the forest, not on the individual trees. Where do we begin in our quest for new thinking? Let's take a lesson from history. As the saying goes, "If you don't learn from history, you're doomed to repeat it."

CREATION OF THE WORKERS' COMPENSATION SYSTEM

As you will recall from Chapter 1, America's workers' compensation system began in the early 1900s in response to problems in the workplace, including dealing with worker injuries. State-by-state, workers' compensation statutes were passed, to a great extent patterned after an earlier federal compensation law for federal workers and the earlier state plans. It was an evolutionary process.

What was the intent, the purpose, of this landmark wave of legislation? This is an important question for us to answer. The intent of workers' compensation was to simply make it easier and, at the same time, more cost-effective for those injured (or worse) on the job to receive benefits, and to standardize what could be expected. This intent was important to both worker and business. It was important to the worker because he or she would not have to go through tort action, and establish work-relation and negligence on the part of the employer. It was important to business because it was more efficient and saved them money. And it was important to both worker and business because it standardized the benefits and removed any doubts regarding benefit to the worker or cost to the employer. The intent of workers' compensation was to help both worker and business.

QUESTIONS FOR TODAY

Let's look at the issues with workers' compensation from an "outside" perspective instead of our usual "inside" perspective. Is today's workers' compensation system still perceived as helping worker and business alike?

Is it still making it easier for those injured or the families of those killed on the job to receive benefits? Let's be fair. Today, in probably 95 percent of the cases handled under our current workers' compensation system, it works. I did not say that it works efficiently. That is another issue we will explore later. The important point is that in at least five percent of the cases, it doesn't. When it doesn't, this is a significant problem to business, the workers' compensation system, and the injured worker or the worker's family. Why? Because it follows one of the Peter Principle theorems. The five percent breakdowns consume 95 percent resources, including time and money on all parties' parts. In other words, five percent of the cases cause high rework rates and administration costs that eat up 95 percent of the available system profits (time and money). This may be okay in government (because we accept inefficiency in government) but in business, it can't be allowed. It would drive a business into failure.

A front-end loader experienced a mechanical failure on an incline and fell upside down into a wastewater pond at the bottom of a hill. The operator was traveling down. Workers seeing the accident quickly went into the water, retrieved the operator and began to provide minor first aid. A fellow operator was driving another front-end loader toward the pond area about the same time the rescuers were going into the pond to help. Jumping from his cab, he began running to the scene to help. He experienced a massive heart attack and died within minutes.

What is the point of this example? It's another one of those five percent breakdowns in the system. The question became: was the family of the operator who had the heart attack due benefits under workers' compensation? Those familiar with the workers' compensation statues are aware that such a case would turn on the question of work-relation. In most states, work-relation requires two aspects—the injury or death must arise out of *and* be in the course of employment[4]. The first part, out of employment, only questions where the accident happened. The second, in the course of employment, concerns whether the worker was doing "work" at the time of injury or death. In this case, the worker had the heart attack on the plant site. The question concerning benefits then focused on the second part. Was the front-end loader operator acting in the course of employment when he suffered his heart attack? Four years later, the case found its way to the State Supreme Court for resolution. The court found that there were other duties that a worker might reasonably be expected to perform arising out of his "work." Responding to emergencies was one of those other duties. His family was, therefore, granted benefits.

Did the workers' compensation system fulfill its original intent—helping both worker and business? The answer is, of course, no, it did not. The eventual end, in this example, could not compensate the worker's family for the "means" it required for resolution. Nor was it in any way helpful to the business or applicable to business in general. Hidden in this example is also the extreme burden such a case places on the system—workers' compensation, business, and worker's family—including time and money. Looking at the intent of workers' compensation, the system failed again.

PROBLEMS WITH THE WORKERS' COMPENSATION PROGRAM

This example emphasizes one of the five problems with the workers' compensation system. These problems include state-to-state differences, the system's lack of competition driver for efficiency, the problems of having a specialized court system, the need to determine work-relation, and the creation of a feeding ground for "opportunists."

---

[4] In some state's workers' compensation statutes, the word "or" is used in place of "and," as shown. This is a significant change in that proof of work-relation becomes much easier, having only to prove one part, not both.

State-to-State Differences

In an era when we are again debating state's rights, as we did in Jefferson's time, what in the world could be wrong with each state having its own workers' compensation program? Let's defuse this emotional argument and look only at the problems it has created.

It's a fact—no two states workers' compensation plans are the same. The statutes are different. What is covered or not covered is different. The rates charged to businesses are different. The benefits are different. The trigger levels are different. The paperwork is different. The time requirements for filing are different. And very importantly, the court interpretations of what the statutes say and mean are very different. One can receive benefits for an injury or disorder if it happened in one state but not if it occurred in the one next door. One can get a huge benefit award in one state and a paltry sum in another.

This state-to-state difference is magnified because each issue or question that requires interpretation is settled in a different state court system. Most often, the statutes only point in a general direction. The court interpretations establish the true boundaries and rules of the system. Each state has different leanings (pro-labor, conservative, ultra-conservative, or liberal) and injects this into their court system. Therefore, even starting with identical language, the interpretations differ widely. In some states walking through employee parking lots at work are excluded from workers' compensation benefits should an injury occur. In others, the same parking lots are included. In still others, it depends on where the security fence or gate is located. Some states extend the work site to include unusual and special hazards a worker is exposed to on the way to and from work such as railroad tracks, hills, etc. Some states exclude claims if there is a pre-existing injury. Most do not, but there are a lot of different tests a re-injury would have to pass, such as lifting limits, exertion levels, etc.

Why is this a problem? If your company only has one facility in one state, you may think that the problem is moot. It isn't. There exists almost no communication between state workers' compensation programs. A worker from another state could enter your workplace seeking only to take advantage of the workers' compensation system and cost you a lot of money. It could be his or her "way of doing business." As a matter of fact, the worker might have done it over and over as he or she moved across the country. You, therefore, become a victim of a scam and end up paying for it. Sound too improbable? It happens all the time!

Additionally, if you happen to have a business in a state that has an overly liberal workers' compensation program, has benefits or costs that are higher than another, or is in financial trouble, those cost you more money. If you have to compete with another business that resides in a more conservative state, has much lower benefits or costs, and is financially healthy, you are at an economic disadvantage. Making your product or delivering your service simply costs you more. Is this fair? Does this make it "easier" for your business, as was the intent of the system in the first place?

What about if you have a multi-state operation, which is common in American business. How does this state-by-state difference affect you? How would you like to run a professional football team in a league in which each stadium has different rules for how the game is played? Anyone involved in dealing with safety, health, or worker compensation issues in a multi-state business knows too well that this imposes high costs on that business and wastes a lot of time. Remember, the intent of the workers' compensation system is to help by making it easier. Wasting time and money is *not* helping. It certainly doesn't make doing business easier.

No Competition Driver

Why would having no competition for a state's workers' compensation system be a problem? Let's ask this question from another viewpoint: Why is the thought of having one company control a particular business, product, or service, a concern to our federal government, particularly the Justice Department? That is called a monopoly. Without competition, that particular company "owns" the market. They can charge what they want.

They can expand or restrict business as they desire. They can make whatever profit margin they wish without worrying about losing their grips on their customers. After all, if they control the market, the price, and the supply. If you don't like it, your only recourse is to go without. Almost everyone is familiar with a board game called Monopoly®. The object is to control the board and eventually run everyone else out of money. What happens when you acquire a second railroad and someone is unfortunate enough to land on one of them? You can charge a higher fee. What happens to the fee if you buy a third or a forth railroad? If landed no, the fee keeps going up, doesn't it? This is more than a game. It is history. A lot of the laws that control monopolies were based on problems that existed in the railroads when they controlled traffic during the industrial revolution. The game of Monopoly® is a fun example of why business monopolies aren't good, especially when you are the one to land on one with your game piece. This is why AT&T was broken up into AT&T and all the "Baby Bells." This is one of the reasons that some mergers between or acquisitions of businesses are not allowed to happen. Monopolies can mean real problems for the consumer.

Additionally, because most areas of the country have legal monopolies for such services as electricity, natural gas, cable television, etc., state and local governments establish regulatory oversight panels. The purpose of these oversight panels is to "balance" the scales in a highly imbalanced market. This is a difficult job. Most consumers have the mistaken impression that these panels are there to protect the consumer from big business. This isn't true. The panel is there to assure mutual benefit—consumer *and* business. This "balance" is what makes the panel's job so difficult. They are continually stuck in the middle, pleasing neither side—a significant problem brought about by having a legal monopoly without competition.

How do these points correspond to problems within the workers' compensation system? Workers' compensation is a monopoly within any state. It is the sole (singular) system for handling work-related injuries, illnesses and fatalities. Generally, it doesn't have an oversight panel to maintain "balance." That usually comes in the form of rate approvals given by the state legislature, a highly political body. It also has the court system to interpret the law or to handle disagreements. Monopolies have specific problems. These are important to our discussion because most are not as obvious as the lack of a competition driver, but result from not having any. These include system efficiency, positive and negative system add-ons, and system control, resulting in a continually sprawling and expanding bureaucracy.

System Efficiency. Competition is crucial for system efficency.

For many years, there was only one manufacturer of a very specialized metal in the United States. This facility met more than 95 percent of the domestic demand. The metal had some applicability to national security so the government was reluctant to allow the use of foreign sources, especially for supplying government contracts. This facility had as near a monopoly as anyone could enjoy in America. Needless to say, the cost of the metal was very high. After all, they could pretty well charge what they wanted as long as it wasn't excessively high. If it was excessively high, it would anger the lawmakers in Washington and could have negative consequences.

Several key people within this facility thought up a much more efficient way of making the metal. With an investment of a few million dollars, they could notably decrease the costs of making the metal. Charging the same price per pound for the metal, the capital necessary to modernize the facility could be quickly repaid and even greater profits enjoyed. Sounded like a good business idea, didn't it? The upper management of the facility wouldn't even entertain the idea. Consequently, the employees who had the idea left the company, acquired risk capital from another large corporation, and built their own plant. Years later, the newer facility had taken the lion's share of the specialty metal's market, placing the older facility at risk of shutting down that part of their business.

What is the important lesson here that applies to workers' compensation programs? The lesson is simple but not easily seen. In monopolies, efficiency and motivation are rare. We are human animals. We need reasons to change and, normally, we need some competitive driver to force us into changing. From texts that detail the dynamics of change, we learn that there are four levels of change triggers. Change triggers are those occurrences that cause us to explore change. Some are powerful triggers. Some are not. Oddly enough, the level of the change trigger translates directly into the level of, energy for, and success of change implementation. These four levels of change triggers are opportunity, need, discomfort, and pain. The first, opportunity, is merely having the knowledge that change is available and that, if taken, it would be good. Opportunity comes without emotion or organizational crisis. For that reason, we seldom change because we have the opportunity to do so. As proven by the upper management of that specialty metals manufacturing plant chose, opportunity is seldom enough. Need begins to get the emotions and organizational future involved. It too, however, is seldom the level at which we choose to change. We are Americans. We need two things to make the decision to change—data and an acceptable level of emotional involvement. Data, meaning we need more than intuition to act. Generally, we need mountains of data, or someone who is a very good salesman together with a reasonably high emotional reason. That's where organizational or personal discomfort or pain becomes a powerful trigger for change.

Monopolies have no emotional reason beyond opportunity or need. That's why change comes so hard to monopolies. They control the market. Why would they change? Change equals risk. When you are risking nothing, why would you risk change? Therefore, for efficiency reasons, monopolies do not seek nor are they attentive to change. Why become more efficient? The market supports the existing inefficiency. Why risk a new expectation on the part of your customers if you became more efficient? In a monopoly, change just doesn't make any sense.

For this reason, monopolies are seldom efficient. Their mere weight is a landmark to their existence. Weight begets power. Power reinforces the monopoly. The monopoly feeds off of inefficiency because there is no acceptable trigger level to change.

What about those who survive working in a monopoly? I say "survive" because that's what they do. There is no reason to be exemplary. The mere baggage of the monopoly drags excellence down. Where is the motivation? Aside from rare exceptions, it doesn't exist. Do you want proof? When was the last time you found a motivated high-performer when you went to the Post Office or the motor vehicle licensing office? How about the last time you dealt with the IRS? In a monopoly, motivation does not exist.

Positive and Negative System Add-Ons. There may be no motivation or reason to change and become efficient in a monopoly, but there is always creativity. In a monopoly, however, the creativity is very often misdirected. Creativity is good. It is the power that drives new and expanding markets. It is the power that makes America great. It is the punch behind our entrepreneurial system. However, it can be a different animal in a monopoly. Why? Because it lacks the very roots that bring about great creative genius. Those necessary roots may be love, hunger, pain, hate, anger, stubbornness, or another high emotional response. Out of these high emotional levels came the NAACP, Apple Computers, McDonald's, movies, books, and our children. What level of emotion exists in a monopoly? What level of directed creativity can we then expect?

That's why I say that creativity in a monopoly is very often misdirected. It comes from the low level emotional roots of perception. In other words, we create things because we perceive that a need exists. This weak root often results in a weaker creative result. An example of this creativity has been the formation and expansion of safety departments (or consultants) in state workers' compensation programs. These state programs are being modeled after private insurance carriers who, among other industrial coverages, offer third party workers' compensation administration or coverage. In private industry, however, they are called "loss control specialists" and have their roots in fire and property protection. These insurance company services have been expanded moderately to provide other specialties such as industrial hygiene services for customers on a cost control basis. It's an industry-wide effort to impact rising business insurance costs.

Now let's be fair. Some state workers' compensation programs have long recognized the importance of preventive strategies as the primary control mechanism in industry for controlling escalating workers' compensation costs. Their purpose in adding and expanding safety consultant staff is to help companies implement safety programs. This is a good thing. However, the number of state programs that have this vision and do this well are very few. Other states have tried another approach, mandating that employers with higher than average workers' compensation rates hire consultants to turn the trend around. This has mixed results. It helps the consultation business but it drives up the costs to the business. Left with nothing else but the mandated requirement, those businesses spend significant amounts when other avenues may be more effective. But a large number of states continue to establish in-house consulting capabilities.

So why are state workers' compensation officers' efforts to enter the safety consulting arena sometimes misdirected? Two reasons—lack of control and efficiency, and uninvited competition with private enterprise. Where insurance carriers may have a regional office of one to four loss protection specialists, they are "lean" and "efficient." As the workers' compensation and loss protection premium rates have become greater and greater, the ranks of supporting specialists in private insurance companies have become more and more lean. After all, these are costs that must be either passed on to the customers or absorbed by the insurance company. Has this same lean and efficient rule be applied to state workers' compensation programs? Quite the opposite. They have expanded. One small western state has more than 12 safety specialists, including an ergonomist and two industrial hygienists. Private insurance carriers that handle near the same volume of business pale in comparison. For example, one carrier has two loss control specialists and another has three in the entire region. Keep in mind that these loss control specialists also have responsibilities outside workers compensation coverage.

Is the cost of all these specialists passed on to their customers? Some times yes, but mostly no. State systems tend to be the highest cost providers (with some notable exceptions). But, in a mandated market with private insurance carriers usually more interested in larger clients who have better than average workers' compensation rates, how can a state program go wrong? The truth is that most state programs are kept afloat with taxpayer money. States started in an effort to control costs just like the private carriers, without equal competition, but only expansion and added costs have resulted, an expansion of the sprawling, growing bureaucracy.

What other impact is this specialized service of state workers' compensation programs having on private industry? In most cases, it is in direct competition with private consulting services. This country was based on a private industrial base, not a governmental base. Does it make any sense to hurt this segment of the private sector?

System Control. Not all monopolies are bad. But one of the areas where monopolies tend to be most dangerous is in the area of controlling the market they monopolize. State workers' compensation programs are no different. Some state programs take this charge very seriously and strive to lead the way to decreased compensation costs to business. Most, however, tend only to control the system. This is done in several ways, including assigning a risk load to private insurance carriers, which has been a widely used tool for balancing the cost problems in some state plans. In other words, having the inside track, many state workers' compensation programs choose to stack the deck rather than assure a level playing field for businesses.

System control also drives up premium costs to business. To counter this, some businesses have been forced to try creative strategies. One of the most interesting is self-insurance. This option, however, has been available only to large companies that could meet the financial assurances required. Recently, small employers have been forming self-insurance pools to collectively meet the financial assurance needed and be able to enjoy some of these advantages. The point is that as state programs try to control the system, businesses try creative ways to circumvent the costly roadblocks that make them less competitive.

The Specialized Court System

Benefit disputes and first-order interpretations of the statutes normally come from a specialized administrative court system that only looks at workers' compensation or other labor issues. It is common for these courts to be comprised of appointed attorneys, who usually specialize in administrative or labor law, and who serve as Administrative Law Judges (ALJ) for some term length. Appeals of decisions made in this administrative court are resolved within each state's court system. However, 98 percent of all disputes and interpretations are made in the specialized administrative court by the appointed ALJ.

This system harbors three problems. First, ALJs tend to be one-sided or more politically driven because of their appointment. Hence, certain ALJs may be known as pro-labor or pro-business by their track record. Others tend to flip-flop back and forth, seeking to "balance" wins on both sides. Unlike with the regular judicial system, which is not as encumbered, many cases result from a "roll of the dice." This dovetails to the second problem: the judgements of this administrative court process tend to be point-specific, not system-specific. As ALJs seek to make decisions in favor of the side they like, a search ensues for a specific point to base the decision upon. In such instances, similar cases are determined differently based on very specific points, not on the greater issues. Decisions such as these become non-appealable and are not precedents for future cases. These decisions become useful to no one, especially not to the system itself. Lastly, decisions made within the specialized administrative courts are only specific to actions within that state. It's very much like reinventing the wheel 50+ times. This isn't efficient.

There were two nearly identical cases in one state. Both involved back injuries. Both involved pre-existing conditions and impairment of the worker due to previous off-the-job back injuries. Both were decided in the ALJ system within six months of each other. The decisions were 180 degrees different. The reason was that one ALJ chose to use a lower standard of legal causation, only requiring that the employee's injury be due to ordinary and usual effort on the job. The award, therefore, went to the employee. In the second case, a different ALJ required a higher standard due to the pre-existing condition and impairment. He required that the injury be due to unusual and extraordinary effort. This employee was denied benefits. Naturally, the last case was appealed on the merits of the first decision. The state's supreme court confirmed the ALJ's non-award of benefits. Did this in any way impact or reverse the first award of benefits? No, it stands as a testimony to the problems with specialized ALJ decisions.

One ALJ is in his third appointment. Ninety-two percent of his decisions have been for the injured worker. Needless to say the insurance carriers don't like this ALJ much and always try to get their cases heard by another ALJ who has a pro-business decision history or another who seems to be more balanced. The state that these particular ALJs reside in is very liberal. It should then be of no surprise to know that both of the other ALJs, the pro-business and the balanced one, did not get reappointed when their terms expired.

The Need to Establish Work-Relation

The need to establish work-relation is a significant problem with any workers' compensation system. It answers the important question of "Do the workers' compensation laws cover this injury?" Let's be candid. Answering this question can be both a curse and a panacea.

As with the earlier example of the front-end loader operator who had a heart attack and died at work, sometimes establishing work-relation can be a curse. Consider the example of the salesman who was hit and killed pulling out of a restaurant's parking lot while going to visit a customer. Was this a work-related fatality? Was the salesman still at lunch or was he in transit to the customer which, of course, would have established that he was at work? Consider another example in which a worker's car was struck by a train and killed. The worker was driving to work on the only paved access road to the facility. The railroad crossing was about two miles from this facility. Did the railroad crossing constitute a special work hazard because all employees were exposed to the same hazard going to and returning from work? If it did, it would establish work-relation. How about another example in which a worker fell to his death at work while "practicing" on his own to become a member of an emergency extrication squad? He was not a member of the squad; he thought that by practicing on real-structures at work after his normally scheduled work shift he would be selected when there was a vacancy on the squad. Did his intention establish work-relation?

The final example concerns a worker who was driving home from work when he fell asleep at the wheel and lost control of his car. It rolled. He was ejected from the vehicle and crushed beneath it. He had just completed his first night's work after rotating to the graveyard shift. Did this rotational shift schedule constitute a special hazard to employees? If it did, accidents on the way home would be considered work-related.

It matters little to our discussion if any of these examples resulted in an award of benefits. What is important is that each was denied at first. Each was pursued through the ALJ system, and one onto the state supreme court. Each took a lot of time and resources to resolve. But resolution of cases like these always have winners and losers. Seldom, if ever, do both parties come away feeling as if they won. "It's a war out there," an attorney for an insurance company said. On analysis, it leaves one with only unanswered questions. Is the system serving the employee or his family by causing the case to drag on for a long time before an uncertain result is rendered? Was that the intention of workers' compensation? Who is left paying for the employee's attorney? Are the costs of attorneys representing business or insurance carriers that very often exceed the potential award justified by precedents established or winning? Does this make any sense at all except to perpetuate the bureaucracy and enrich employee attorneys?

How could establishing work-relation be considered a panacea? To many employees, establishing work-relation is the difference between having dependable benefits (often for extended periods) and not bringing in any income and not having the medical bills paid for. This, you will admit, is a big difference.

Please don't assume that all employees file claims to take advantage of unsuspecting employers because they want a free ride. Many workers' compensation claims are legitimate. In many cases, especially with lower to moderate wage workers or those who cannot afford or do not have medical or extended income protection, abuses occur. These are not bad people, only resourceful ones who care about feeding their children and keeping their houses. Suspicious injuries and illnesses that fall into this category include back strains, hernias, hearing loss, carpal tunnel syndrome and a rising number of other syndromes. "I don't know, Doc. I just bent over to pick up my tools at work and now I can't stand the pain." Work-relation *is* established. Now, Mr. Employer, it's your job to disprove that the injury occurred. This is not a level playing field. Any employer that has gone though this knows this all too well. But it is the workers' compensation system that has created this. Establishing work-relation is a significant problem from all perspectives—employee, employer, and compensation carrier.

A Feeding Ground for "Opportunists"

We've fallen into a deepening spiral in this country. Philip Howard talks about it in his book *The Death of Common Sense*. In an effort to get away from following the "law" of common sense in this country, we have tried to define everything as a matter of law. In other words, instead of saying, "Mr. Employer, you have the responsibility to make your workplace as safe and healthful as you possibly can," and allowing any challenges to be measured against "reasonableness" and "good faith," we have tried to define, most times in great detail, what a safe and healthful workplace is.

This is the same approach that we have taken in workers' compensation statutes. The purpose of this approach is to not have to answer the basic questions all the time. Instead, we have traded the basic questions for answering all the detailed questions, of which there are many more. This spiral causes us to continually deepen it by adding to the regulatory details. This, of course, begets even more detailed regulations. The cycle continues, *ad nauseam.*

What is the root of these increasingly detailed regulations? Who asks or causes us to ask for more detailed regulations? In a word, "opportunists." An "opportunist" is a person who pushes the envelope, finds loopholes, and takes advantage of them. I find it is difficult to distinguish between the words "entrepreneur" and "opportunist." Both "go boldly where others fear to tread." Both take considerable risks. Both have the potential to gain a great deal. Both are oddly American concepts. In any case, one cannot argue that the continual parade of opportunists that find ways to take advantage of loopholes or flaws in the workers' compensation system is not a considerable problem.

ZOOMING IN OR BACKING AWAY FOR CLARITY

A common theory is that there are a lot of "evil" employees who would cheat the system given the opportunity. Personally, I don't believe this to be true. I'm an optimist about people. I believe that people are basically good and honest (emphasis on the word "basically"). I also believe that there is a certain percentage of basically good and honest people that will take advantage of "opportunities" should any system provide them. Call it "entrepreneurial spirit" or "going for a quick buck," whichever you prefer. But there are those who, given the flaw in the system, will find it and will take advantage of it. Does this make them bad? Heavens no! It only makes them opportunists. Is the answer to continue the same path we have traditionally taken—add more detailed regulations? No, that has never worked. We've always had the wrong focus. The opportunists *aren't* the problem. We've created the flaws within the system that give them the opportunities.

Those aspiring to the "evil" employees theory tend to play both sides of the fence. They also argue that there is no way that all the flaws in any system could be removed. This is true from any viewpoint, although we repeatedly attempt to plug all the holes by additional regulatory detail. Detail is a reductionistic way of thinking. No solution can be found in this direction. Our history should teach us enough to convince us of this fact. On our way to understanding, however, let's choose another path.

Rather than dissecting all the different workers' compensation programs looking for these flaws which allow "opportunists" a foot in the door, let's deal with the root cause of the problem with workers' compensation. One that encompasses *all* of our identified problems with the workers' compensation system. What is this root cause? Simply stated, workers' compensation is an isolated and specialized system. What is the problem with having an isolated and specialized system for handling workers' compensation? After all, it is the way we deal with almost everything in our society. For example, we determine that we have an environmental problem; the Environmental Protection Agency is created. In order to solve the problem with worker safety and health, we built the Occupational Safety and Health Administration. Society needed to become outraged about drunk driving in this country; Mothers Against Drunk Drivers (MADD) evolved. If it was determined that more national attention was needed to save the Spotted North Fork Plains Boll Weevil, we'd start a special interest group. It's a fact. This is the way we always deal with problems, or perceived problems, in our society. We identify a problem, then we create an isolated specialized system or group to "solve" it. In the long run, does this approach work? Of course, it doesn't. It never has.

Why hasn't this approach worked? The specialized and isolated focus inevitably causes a condition called "Cause-Myopia." Cause-Myopia has a number of results. First, the system or group becomes so focused that they can only see their perspective. It becomes THE perspective. In any society, the trick of living together lies within our ability to balance perspectives. Incidentally, that's where Hitler, Stalin, the Klu Klux Klan, and many others went wrong—they had no balance. It was always "My way or the highway."

Second, Cause-Myopia results in a continual effort to define everything within THE perspective. Call it the "deepening spiral." The rules get longer and more detailed. Everything becomes more and more formal and requires multiple approvals for it to be official or "blessed" as representing THE perspective. It's really a function of not having any trust or faith in common sense. When those in power who announce that they are going to simplify government, they really don't stand a chance. This is the second result of Cause-Myopia; flaws are found in every isolated and specialized system that opportunists seek to find and take advantage of.

CONFOUNDING THE PROBLEM

This problem of being isolated and specialized has existed from the beginning of the workers' compensation programs in America. But, like few other systems, workers' compensation has a significant confounder. Every state, territory, etc. has its own workers' compensation system. None of them are the same. So, America doesn't have one large problem with workers' compensation, we have more than fifty of them. Because of this, it's futile to climb on the "fix the system" bandwagon. First of all, there are too many workers' compensation systems to fix. Second, you can't fix any system without fixing what caused the flaws in the first place, the root cause.

Doesn't it seem like "you can't get there from here"? In truth, the solution really isn't as complicated as the present day quagmire. The solution focuses on the root cause of the problems with the fifty-plus workers' compensation systems. It focuses on the isolated and specialized concept from which workers' compensation was derived. Let's consider a few parallels to put this into context. Would it make any sense to have isolated and specialized places to fix cars? To some extent we have this now, after all, it's the way our society thinks. But let's take it to the extreme. Let's have a shop that fixes fuel injection systems only. Another shop repairs automobile electrical systems. A third shop works on auto exhaust systems. If you needed a tuneup for your car, you'd have to visit all three shops (assuming that each could do their job independently) to get what you want—a smoother running car.

Let's say that medical doctors became even more specialized. If you got the flu, you'd have to visit a flu specialist. It doesn't make any sense does it?

Why then do we think it makes any more sense to have specialized systems to handle workers' compensation, medical, and other problems that are isolated by cause or age? For example, if you get hurt at work, the workers' compensation system handles it with its own rules, paperwork and associated flaws. If you get injured at home or in your car, your medical insurance takes care of it with its own set of rules, paperwork and flaws. If a guest gets injured at your house, your homeowner's insurance steps in with its own set of rules, paperwork and flaws. If your son or daughter gets injured at school, the school's insurance coverage steps in with their specialized rules, paperwork, and flaws. If you happen to be in your "golden years," Medicare and Medicaid add a whole new dimension of rules, paperwork and system flaws. What is the number one problem all of these different systems encounters? People taking advantage of them. Is it any wonder that there are so many people who take advantage of the different medical care systems? No, the answer is *not* to fix the workers' compensation system. The answer is to stop this ludicrous way of thinking! If we are going to have any system that works, we must simplify the quagmire, not add more rules and paperwork to it.

FIXING THE SYSTEM

Medical Care Administration

Let's ask a simple question. What is the basic purpose of Medicare and Medicaid, medical insurance or workers' compensation? Isn't it to provide the necessary care, medical or otherwise, that is needed? What separates these systems? It is a variable such as age, employment, place of injury or illness, etc. If we are concerned with providing the necessary care for people, do any of these details really matter? Remove yourself from our traditional specialized-system pattern of thinking. Mull this question over from a common sense vantage point. What makes the most sense, knowing the purpose of these different systems? Do the details concerning age, employment, place of injury or illness really matter? If the purpose of providing care for people is valid, we have only one way to fix the system. Fix the entire medical care administrative system.

Am I calling for elimination of the workers' compensation system? Yes, as a matter of fact, I am. There is no need for a specialized workers' compensation program, and other specialized programs for that matter, which are an outdated outgrowth of our traditional ways of thinking and have moved away from their intended purpose. Other systems that are designed more efficiently to provide administration over medical or other services can be easily converted. The only issue, of course, is who will pay for the costs of work-related cases. This is an accounting function only. It makes no sense to have a specialized system (that does the same things other systems do) that is only distinguished by origin of injury or illness.

Returning to our focus, would fixing the medical care administrative system help improve worker safety and health? Yes, it would. It would have several notable impacts. First, it would make the system less apt to be "played with." Most of the problems that surround any of our present specialized medical systems focus on inclusion or exclusion within that particular program. That's where we have spent a great deal of our time, perhaps 90 percent, writing details within the statutes. Defining who is in, who is out, what is in and what is out. Simplifying *one* system with only inclusionary language, and that being correspondingly simplified, doesn't invite opportunists. After all, everyone is in. By simplifying, you place a sign that says, "No opportunists need apply."

Second, reduced systems equals reduced system administration. Reduced costs can make business more competitive and place a smaller burden on our society. These savings could be put back into correcting workplace hazards or into the training of management and workers. In short, simplified systems save money.

Third, preventive strategies have a much greater impact when a holistic approach is taken. For example, why redesign a tool a worker uses when redesigning the workstation would be more efficient in the long run? It's a long-term approach instead of a short-term one. Now realize that this is counter to the way we in American business are accustomed to thinking. Traditionally, we put bandaids on things because we only focus on the next balance sheet, like the next quarter's financials. This long-term concept is more Asian in its thinking—Japanese, to be exact. The Japanese have a word for it. They call it *poke yoke*, or fixing a problem so that it can never happen again. It requires a complete focus on the problem, wider attention than just a specific instance of the problem, lots of participation, and never accepting bandaids.

Fourth, system simplicity reduces the stress on workers concerning what plan certain conditions fall under, or what paperwork is required, or what coverage is lacking. Medical care is a big worry. After all, that was workers' compensation intended purpose: to provide assurance that if an injury, illness, or death resulted from a work-related cause, the worker or his family would be taken care of. No worries, only assurance.

The Changing Times

The workers' compensation system as it now exits is not bad. You probably think that I don't like it. Wrong. If everything is left as is in our total system for worker safety and health, the workers' compensation system is fine. But it's hard to be happy with the present state of worker safety and health in America. It is going nowhere. It's reprehensible. If we accept America's position in the world economy, the present workers' compensation system is also fine. After all, it is symptomatic of the bureaucracy that strangles advancement in this global market. It's hard to be happy with the position of America's businesses in the world. This means too much to our children and their children. If we believe that nothing can fix our overall governmental system, our present day workers' compensation system is fine. If we believe that there is no way to provide efficient, barrier-free, easy-to-use injury or illness benefits for America's workers, the present system is fine. But having worked with the system for many years, we cannot capitulate to the idea that efficiency, barrier freedom and ease cannot be achieved.

We can live with our present workers' compensation system. Millions have for many years. But times change. Economies change. Economic position in the world changes. Our perspectives change. Our patience changes. How long are we willing to "live" with the present workers' compensation system, which many of us deal with daily, and not demand change? Why is total change warranted? Because we were inattentive years ago when it stopped functioning as intended. Oh, we weren't alone. Generations have passively "lived" with the mistaken impression that government had a major role in taking care of us. Passively, we allowed it to grow and grow and grow. Efficiency, the nation's budget, and the intent of the systems were the victims of our inattention.

The system stopped accomplishing its intent years ago. We became too caught up in the details. We worried about "opportunists." We became frustrated over state-to-state differences. We dealt with the politics of the ALJ system. We lost sight of the intent of the workers' compensation system and how our economy, country and businesses had changed. We simply got too caught up in the details, thinking that they could make everything okay. Now we know that they cannot.

Earlier, you were asked to move away from our natural tendency to look traditionally at workers' compensation. This is not a traditional approach. Begin with a "what if . . . " approach, as a child not knowing any better would. That's the way our forefathers thought when they drafted the words that guide this republic. Why is it that we seem to have lost the ability to think independently and see a different tomorrow?

SUMMARY

Current problems—what we should stop doing:

■ Dealing with work-related injuries and illnesses as a specialized system.
■ Accepting the inefficiencies and ineffectiveness of the workers' compensations programs as "the best that we can have."
■ Accepting a legal monopoly that drives up costs to business.

Fixing the system—what we should be doing:

■ Questioning if a specialized system for worker-related injuries and illness is still valid in our changing business environment and society.
■ Questioning if de-regulating the workers' compensation programs is viable.
■ Demanding efficiency and oversight for workers' compensation programs.

5

THE INEFFECTIVE REGULATORY ENVIRONMENT

I remember wearing a full-face respirator hooked to an acid gas/carbon monoxide canister and climbing to the top of an experimental chlorinator to take off-gas samples. This was a normal job that new recruits had to do. I drew the short straw. Below me were freethinking engineers trying (I hoped) to keep the molten salt inside the vessel and piping. Too often they didn't. Too often, I retreated down the ladder from the chlorinator's top due to a "slight" breakthrough, and I found my CO color indicator on the canister "past due." I must admit I became callous to it. Unfortunately, so was my management. It happened too often. In fact, it was always seen as a joke to them.

Then was the time when molten salt melted through a piping elbow on the chlorinator. As it began to uncontrollably gush from this newly created orifice, it ran slowly toward an open water collection drain. We ran in the opposite direction. If the flow was successful, molten salt and water would be an explosive combination. Fortunately, the salt flow didn't make it. Those who were left curiously watching or frozen panic stricken should have been very glad it didn't.

The number of us who remember professional life before OSHA is getting smaller. It gives a lot of perspective, being able to compare today, yesterday, and the path we've traveled. None of us would have thought it would be as is it is today. Some would have thought better, some worse. But none of us would have been able to gaze into our internal crystal ball and say, "Yep, that's exactly the way I thought it would turn out." Back when I was a young industrial hygienist, I feared the worse. An out-of-control monster, OSHA, would bring the mighty sword of "thou shalt" down on industry, who was doing nothing but "thou shalt do what you wish."

I remember walking across the plant toward the main gate to go home one day. It was dusk and the sun had just settled beyond the horizon. I noted a different hue in the air just ahead in my path. The dimming daylight made the color and detail hard to see. Was it something or nothing? Being trained by experience to be cautious (my advantage), I stopped and moved to another angle to get a better look. There was indeed something lying in my path. It was a cloud. The cloud was green. These signs indicated chlorine, probably concentrated at a dangerous level. Five minutes later, when the light wouldn't have been as good, I would have walked unknowingly into that cloud. I doubt I could have gotten a second breath. Someone else would be writing this book then.

FRIEND OR FOE?

You might think the two examples above are unimportant examples. Let me ask a question. Before OSHA, what recourse did we have? I hate when someone else irresponsibly messes with my chance to have a tomorrow. It really irritates me. What recourse did I have then? Complain to my supervisor—the same supervisor who thought it humorous that my CO indicator showed my canister was spent? To his credit, he was only a product of his American industrial training and upbringing. Those of us who grew up post-OSHA, or have always been in the staff orientation, may have a hard time warming up to these examples. Trade places in your mind for a moment with a worker-bee prior to OSHA. Think about it. It is too easy for us to "lick our wounds" dealing with the here and now, the reality of OSHA, and never put ourselves in the worker's shoes prior to OSHA's creation. For our purpose here, it's an important perspective to begin with. Otherwise, we might skip over the messages that lie between the lines in this discussion.

Hindsight is, of course, always 20/20. In the early 1970s, the thought of OSHA was more than frightening to industry. Keep in mind that the EPA was also in its infancy. Industry was only just beginning to feel the potential impact of the EPA. With the possibility of a second part of government, OSHA, looming before them, industry feared that government was going to run industry out of business. OSHA singlehandedly was going to make it impossible to make a profit. The agency just wouldn't understand. They would be the enemy, the friend of organized labor.

Have you noted a certain level of paranoia? Regardless of who you are or where you come from, OSHA tends to do that to you even today. One minute you welcome their entry. The next minute, they are idiots enforcing idiotic laws that don't make any sense. Does this sound familiar? We have all been there. Not once, not twice, but countless times. Industry, government or labor, it doesn't matter, everyone has a love-hate relationship with OSHA.

This whole phenomenon probably comes from the fact that OSHA and the other agencies created by the Williams-Stieger Occupational Safety and Health Act are complex. It isn't a simple matter of like or dislike. Some parts you like, even love. Other parts, however, you loathe. It isn't an easy puzzle. But, to get to into our discussion of the regulatory environment, we need to dissect this puzzle, piece by piece.

Let's start unraveling this puzzle by asking and answering a basic question, a question that has been widely asked and debated for almost thirty years: Has the existence of OSHA helped worker safety and health? From an insider's point of view—one that saw industry before and after OSHA, one that has worked in government and industry, and one that has "lived" through the paranoia and double-faces OSHA creates—YES, OSHA has helped worker safety and health. Now before you stop reading and conclude that the final word about OSHA is that it is a positive part of the system for worker safety and health, let's ask and answer another question. This is probably more important today than the first question. Has OSHA continually improved worker safety and health in America? After all, that was OSHA's charge. The answer to this one, is of course, NO. The difference between these two questions is significant. The first is frozen in time, comparing yesteryear only to a snapshot of today. You simply could not answer that question any differently than yes. The second question looks at the continuum, not one point against another. It looks at improvement, yesterday compared with the next day, then the next, and the next, until you reach today. This is an important difference. Because at the root of this question is the exploration of the problems that have limited the mission and impact of OSHA and the possible solutions in our quest for improving worker safety and health.

FROM GENUINE CONCERN . . .

OSHA wasn't created just to make industry safer. That point may have been pushed a lot by supporters, but it really wasn't the point. The Williams-Steiger Occupational Safety and Health Act was an attempt to improve worker safety and health. It was a genuine concern for the human element, America's workforce. You might say, "Aren't they the same thing?" No, they are definitely different. Think in terms of automobile safety. If you focus on the automobile as the key component of safety, you would require five-mile-per-hour bumpers, collision panels in the door, air bags, seat belts, and look at automobile collision research. This is a "thing" focus. This focus on the automobile is like a focus on industry. If you focus on a genuine concern for those driving and riding in automobiles, you focus your program differently. You establish safe driving speeds, require the use of safety belts, make laws about using turn signals to change lanes or make turns, etc. This is an "act" focus. This is like a focus on workers and management.

There is an important lesson for us here. First, isn't it apparent at this point how the legislation and the regulations went array? Having a genuine concern for workers' safety and health, Congress focused on industry "things," like the automobile in our example. They forgot to hold the workers responsible for "acts" that compromised their safety. This lesson is unlike the approach that traffic safety focuses on "things" a car must have, can never be successful. It is essential that there remain a strong focus on "acts" of those who drive automobiles. How is this different from OSHA's focus? It isn't.

More Than OSHA

OSHA wasn't the only thing that was created for the purpose of improving worker safety and health. Actually, four different entities were created by the Act. OSHA is the most widely known because it got the policeman's hat. They were the enforcers. Because labor was the focus, not industry, OSHA was placed in the Department of Labor, NOT the Commerce Department. But what would OSHA enforce? The Act made provisions for some rather hastily put together, hodgepodge regulations. That was a disaster. But the Act looked to the future also. It created NIOSH (National Institute for Occupational Safety and Health) with two charges. NIOSH was to research and recommend standards for OSHA to put into the regulations and it was to lead the way by training practitioners in the practice of safety and health. This agency was placed into the health side of the government. Congress had great foresight in making NIOSH part of the Act. The Act also created a special courts system, the Occupational Safety and Health Review Commission (OSHRC) as a part of the Justice Department and an Advisory Council (that has never worked) that would somehow orchestrate the efforts for improving workers' safety and health.

There was significant justification for the Act. First, the regulations would provide a single reference for important safety and health issues. Guidelines couldn't provide this. There were and are differences of opinion. The regulations would provide one voice, a definitive one that would have the power of law behind it. No longer would it be a "thou should" as it was with the recommended guidelines. It would be a "thou shalt," enforceable by law. Third, it would be the intent to police industry to make sure they were doing what the new laws said they must do. It was a trust issue. It's the same trust issue that puts police cars on our roads to see if we obey the traffic laws. It was a trust issue, plain and simple. You tend to trust what your eyes tell you, not what you are told. Fourth, OSHA and NIOSH would serve as an avenue for employee concerns. Before the Act, employees really didn't have much recourse if they had safety and health concerns that their employer didn't agree with. Arguably, they did if they were organized and had a contract. But organized labor has never represented the majority of American workers. And fifth, there would be a significant improvement in the education of practitioners of safety and health, and needed research on unknown health impacts. This would provide for improving worker safety and health in tomorrow's workplace.

Who supported the Act? Naturally, organized labor did. Industry was surprisingly neutral. Some came out against the Act while others seemingly supported it. It was a political stance. If industry killed the Act, they might get a worse one or be branded as anti-safety and health. Either way, these were negative. Also, if they semi-supported it, perhaps OSHA would be kinder, not as angry with industry, as if they had vehemently opposed it. This is commonly called playing both ends toward the middle. It was politics at its best.

So, the Act was passed and signed into law by President Nixon. The gears were put into motion to create the initial regulations. Staffing of the new agency was started. NIOSH began its infancy drawing from the Public Health system already in place. An advisory board was begun, but sputtered as it does today. The search was on for staff for the new Review Commission. Everything the legislators put into writing began taking shape.

STATE OSHA PROGRAMS

Knowing that some states opposed federal control, a states rights issue, language was adopted that allowed the creation of a state-focused OSHA function. There were only three catches for the states' independence: they would have to shoulder the majority of the costs of the state OSHA program, they would have to enact similar regulations that were at least as stringent as the federal standards, and they would have to be happy to have federal OSHA look over their shoulder constantly. For many states, this was a love-hate relationship. The state had the mandated option of having either the feds come in and usurp their rights, or paying the price to have their own program. They hated having an "as stringent" straight-jacket placed about them and having "Big Brother" always pointing out when they screwed up. The feds loved having the mandate and the power over the states. They hated not being given the whole pie to eat. It was a power and turf issue.

Having states take the option of running their own OSHA programs created the same problem as workers compensation. No state programs or regulations were the same. The state OSHA programs differed widely. Others were staffed marginally. Some were staffed twice as well as federal OSHA was. Some states copied the federal standards word for word. Others added words, sentences, paragraphs, pages, and volumes. As a consequence, federal OSHA hated some states because they were inadequate or just copy cats. Others were despised because the feds couldn't keep up. It was a competition thing. To combat those states that, in OSHA's opinion were inadequate, benchmark staffing levels were established and an attempt was made to ram them down the states' throats. You can't push a mule through the mud. The only way is to lead and entice him. OSHA never got that lesson. California OSHA (Cal OSH) is an excellent example of a state program that left federal OSHA in the dust. Cal OSHA didn't just get religion, they became evangelical. They had specialists and staffing federal OSHA couldn't even dream of having. They never stopped at the federal OSHA regulation level; they wrote and enacted more rules where federal OSHA couldn't. It was, as one CEO of a California facility put it, "a feeding frenzy. The only problem was that there wasn't enough industry to feed on. So, they went back for seconds . . . then thirds . . . then fourths."

NIOSH: A DIFFERENT BEAST

One of the charges given NIOSH was to provide research and information on the need for new regulations. This research took many shapes. One of the most interesting is the Health Hazard Evaluation or HHE. In this program, NIOSH conducts "research" at specific industrial facility locations. These HHEs can be initiated by statistics and other research findings, as part of an ongoing HHE, or by an employee request. In any event, depending on what NIOSH "feels" is an appropriate level of research, a HHE can take minutes over the phone to years or complete, involve one or two individuals or hundreds. In other words, it can be no big deal to a facility, or a major business interruption and cost.

There are a few problems associated with HHEs that are important to discuss because they are very different from OSHA inspections.

1) There is almost a sacred belief placed on HHEs. Therefore, in an effort to chase the holy grail of research, there are not the normal safeguards to industry that are provided in OSHA inspections. Like OSHA, NIOSH can get a warrant for entry, but unlike OSHA, NIOSH doesn't have to limit the focus. They can look at what they want, wherever they want, for as long as they want. Let's face it. Researchers like to do research. That's why they chose that as their occupation. And, once in the door, their love of doing research tends to transcend rational or practical judgement.

2) HHEs tend to become "fishing expeditions." After all, if NIOSH finds nothing, it doesn't look good for the HHE staff nor for NIOSH's research needs. So, they tend to look until they find something and then continue to pick at it until they have some kind of research finding. In other words, if NIOSH wants to come into your facility, they *are* going to find something.

3) If OSHA inspects your worksite, whether or not a citation is issued, it always stays in your file at OSHA. Those files, in reality, have limited public access. The result of an HHE is a published report which is available to anyone who wants a copy. In other words, HHEs can be *very* public. Because it's research, you don't have any say about what goes into

or is omitted from the report. Consequently, highly proprietary information, innuendos, unproven or unsupported health effect allegations and other highly damaging things can be placed in the report for everyone to read.

Okay, let's be totally fair. Copies of these HHEs are not available on every corner. You have to know where to look to find them. But, the point is, if you know where to look or ask the right people, the information is available to anyone.

COMPLEXITIES OF OSHA

OSHA was not created a simple beast. It was indeed complex. The interesting part about complex things is that they increase in complexity as time goes on. OSHA has certainly not been different. Beginning at zero, everything was new and created for the first time. That was a one time occurrence. In a governmental system, a dichotomy is created—those who are here today and gone tomorrow and those who are always here. In industry, job security is solidly wedded to continued value to the organization. That's why terms like RIF (reduction in force) and downsizing are popular. This has *never* been the case in government. Once you are hired, it seems that you are entitled to a job until retirement. Look at those who drafted the working documentation, the Congress. Where are they today? There are two camps within any government agency. We call them the political appointees and the bureaucrats, those who will only be there while the current party is in power and those who will remain regardless of the party.

Why is this important in our discussion of the complexities of OSHA? Because there are two coexisting yet interdependent camps. The political appointees are the ones that make headlines about the direction OSHA is pursuing or defending. They allegedly plot the course for the agency. This, of course is not totally true. It is the bureaucracy, the long-term employees of the agency that make the day-to-day operations work and determine the real direction of the agency. Why? Because they don't go away with the new administration.

This is both good and bad. Good, because you cannot imagine the confusion and lack of direction of everything in the government if it truly changed direction every time the ruling party changed. Bad because change in direction takes much longer than any one party has to really change what is going on or where that agency is going. The further complication is that as time goes on, it gets more and more difficult, given the longevity and difficulty, to change the true bureaucracy.

This reality has created some interesting problems in OSHA. For example, the front office, the political appointees, might say that OSHA was going in one direction when, in reality, the mechanism, the bureaucracy was going in another. Once the machinery of the bureaucracy gets moving in one direction, it is extremely hard to turn it. The thought process of long-term OSHA employees is that this is merely the administration of the moment. The next administration will probably see differently than the current one. Why change for the short-term? So the machinery continues on, not efficiently nor at a fast pace, but just on.

A "CLUB" AGAINST INDUSTRY

The Performance Standard Is Born

So, in the beginning, what were the complexities of OSHA? Keep in mind that many things were unexplored territory. OSHA was originally created to be a "club" against industry. The word *club* can be taken in the most literal definition. This was unexplored turf. Very quickly, however, an Idaho case, *The United States of America vs. Barlow*, quickly defined OSHA's power. They could indeed run roughshod, over industry having the warrant power to enter any place of business. OSHA could indeed be used as a club against industry.

In the beginning, OSHA took a chapter-and-verse approach. What OSHA writes down in the regulations, you, Mr. Employer, must comply with. This turned out to be a two-edged sword. It was good from a compliance policing position. Forty-two inches was 42 inches. It was a black and white world. It didn't take any thought on the part of the OSHA inspector to gauge compliance or issue a citation. But, Congress is a political body. Business has a large voice in Congress because it provides a great deal of money to the legislators' re-election coffers. Money buys influence. Get the connection? Business began to whine a lot about the verbatim compliance issues enforced by OSHA.

The old adage says, "Having is not as good as wanting." Industry thought that if there were fewer "thou shalts," that if they were replaced with directional language, it would make compliance easier. In other words, don't tell us how to accomplish it, just tell us what the desired final result should be. The performance standard was born.

Hazard Communication Standard

The first performance standard was the Hazard Communication standard. In reality, having only to write a performance standard was a relief to OSHA because they really didn't know how to write a "thou shalt" standard of this magnitude, dealing with such a dynamic area as chemicals and chemical products. This performance standard became, and continues to be, to this day, an albatross about the necks of industry. Let's look at two points to make this very clear. First, what triggers noncompliance with a performance-based standard? It isn't an OSHA inspector's opinion. It is a complaint or an exposure. The logic goes something like this: If the employee complains or someone gets exposed to a chemical, obviously you don't meet the performance expectation of the standard. Try fighting that one in a review commission hearing or in a court of law. You can't win it. Secondly, point to any one system or program that is used universally throughout industry for compliance with the labeling requirements of the standard. You can't because there have to be at least a million of them. In fact, *nothing* is standardized. This has caused companies to invent their own programs, and incur the costs of doing so, more than any other compliance aspect. There have been countless business dollars wasted creating custom labeling systems, and employees are confused because their new employer's system does not look anything like the one at the last place they worked. An industry cannot live with these wastes and errors. If only they had a "thou shalt," verbatim-type standard like they didn't want, they could have saved hundreds of millions of dollars.

This is not a call for verbatim standards. It is only being used as an example of the complexities associated with the creation of OSHA.

Complexities from Creativity and New Administrations

We like creativity. But an agency left in a state of boredom or the influence of a new administration can have both positive and negative creative impacts. These are significant complications on what we know and expect of OSHA. One of the most notable is the voluntary protection program, or the VPP. Originally created in Reagan's kinder and friendlier government style, the VPP was an attempt to change OSHA from an aggressive, antagonistic enemy of industry to a partner for worker safety and health. Admittedly, this was more in line with the original intent of the Act, to improve worker safety and health, not to beat up on industry. But, as it came out to industry it looked more like a Trojan Horse. This is not meant to create a negative image of the VPP; in fact, it has had a lot of positive impact for workers safety and health, getting labor and management to work together. But its impact wasn't immediate, was it? It took time because it was so "different." Most programs that have "appeared" in the OSHA plan haven't been as successful. They've become albatrosses of the system, not fitting the style or culture of OSHA. They were a lot like a coal miner trying on silk gloves. It just didn't feel right to OSHA nor look right to industry.

Take a mid-1990s example. When Assistant Secretary of Labor Joe Dear took the helm, he verbally reversed the momentum of the VPP. He didn't like it. He wanted his own, meaning the current administration's, version. Understand that Joe Dear came from the state of Washington, which doesn't have a VPP. So immediately the administrative direction and verbiage changed. It was extremely poor timing in that the VPP was really taking off. Admittedly, administration of the VPP takes a lot of time away from the core mind set of OSHA, compliance. The shame was that the new administration didn't even research the VPP to see if it had merit or was successful. It was devised by a Republican president. It couldn't have value. What happened? Eighteen months later, Joe Dear said he loved the VPP and was going to use it as a major thrust in his administration. I'm sure that the change in congressional control from Democratic to Republican had nothing to do with this change of heart.

Looking at this simple example, what complications did it create for workers' safety and health? Although there are no statistics or polls to prove this, let me share some opinions. First, there were some companies that stalled their move toward the VPP because they weren't sure it would be there. State OSH programs that were moving toward VPP suddenly de-emphasized their program, not wanting to lose favor with Big Brother. This represented another politically-motivated stall in advancing worker safety and health. The machinery of OSHA changed the focus away from supporting VPP and conducting evaluations, because that was not in the current administration's favor. The system, and the state programs, became inefficient waiting for the current administration to make a long-term decision. Was this simple example a significant complication to OSHA's intent of improving worker safety and health? If it slowed progress in any company or caused one employee to be injured or killed, yes, it was significant. Politics and ignorance don't help. They only complicate things.

Bigger "Clubs" Make Industry Comply

Let's look at another notable complication at OSHA, the egregious multiplier. This is an excellent example of a solution which not only did not fix the system but added more antagonism and problems than it solved. In reality, the existing penalty structure is minimal. It just isn't enough to get the financial attention of some businesses. But is increasing penalties the right direction to follow? We've already stated that changing the system is the only way, but let's follow this particular line of thought. Conventional thinking says that enforcement is the only way, and high penalties are the only path to making industry comply. So the thought behind the egregious multiplier was that if a company appeared not to be making an effort, OSHA should count each infraction and multiply it by a much higher number. This made penalties easily jump from a few thousand dollars to millions. Did it have its desired effect? Let's ask another question in response. How many of the multimillion dollar penalties were contested and ended up being settled for much less, even the penalty amount if the egregious multiplier was not used? The answer is, most of them. Did this complication, the egregious multiplier work? Well, it employed more attorneys, for sure.

During the last Democratically-controlled Congress, there was a move to include criminal penalties in the OSH Act. Could this have been a significant complication to OSHA? Was the egregious multiplier a big complication? And, the criminal penalties didn't even talk about jail time. Why do we continue this lunatic way of thinking that the club is the only way? Hasn't our experience taught us anything? Hasn't the success of the VPP whispered anything in our ears?

The Inspection Game

I grew up in the old school. OSHA was the enemy. I could say that I was naive or influenced by others. The truth, however, is that I just didn't question it. Doesn't make me very smart, does it? Inspectors showing up at my gate were treated with stern a response, a letter-of-law approach, and strict limitation of inspection focus. The industrial representative's communication would be almost clinical. In other words, it wasn't a warm reception. It was common for the inspection to be interrupted by consultations with corporate or OSHA administration conferences to resolve or negotiate a particular point or OSHA request. Everything was provided just short of demanding a warrant from OSHA.

Over the past twenty years, however, my approach has changed from this reactive and combative approach to one that is proactive and cooperative. I admit it; I'm a slow learner. I recognized that the intent and success of the workplace system for worker safety and health depend on both OSHA and industry working together, not against each other. The only thing that gets in the way of this is our traditional approach toward one another which is, of course, purely a product of our paradigms. This can and must be changed.

Working with a lot of different companies, however, I find that the first approach, the combative approach is much more common in American workplaces. Perhaps this is as common as 95 percent. The only variable in this is the degree of outward antagonism displayed during an inspection, ranging from open to subliminal. This combative approach to OSHA inspections has begotten a "game" that industry plays. Call it the Inspection Game. Like most games, it has winners and losers, and an objective and duration. The duration spans the inspection from requested entry to resolution of all issues. The objective is the most complex element of this game. OSHA usually knows the regulations better than most industrial representatives and having seen a lot of other sites, they know where to look and the questions to ask to find violations. The company knows the particular site better and where the problems are so they can "steer" the inspection and "hide" information to a greater extent.

One of my favorite aspects of this Inspection Game is a common practice known as "baiting." A compressed cylinder is purposely left unsecured. A grating is left off in a common walkway. An extinguisher is blocked with a trash can. A minor infraction is purposely placed in the path the inspector will be led along so that it can be easily found. I've been on inspections where the inspector's attention is drawn to another issue and the guide "points" out the infraction with something like, "Oh great, I was just down here and look at this." The purpose of baiting accomplishes two objectives of the game. It allows the inspector to find infractions so that he or she doesn't have to dig too deep and it also tends to shorten inspections. You might find this scenario to be a humorous and isolated instance. In workplaces that believe the traditional "OSHA is the enemy" paradigm, however, this game is common. It is practiced in the majority of American workplace inspections.

Differing Perspectives

The Inspection Game is, of course, driven by different perspectives. OSHA inspections follow traditional patterns. Traditionally, an inspection is *not* conducted to improve worker safety and health; it is done to find citable violations. It is a boggy system that is measured and used to measure the effectiveness of the OSHA program, and to seek funding. Ask yourself, what would happen if inspections suddenly focused on improving worker safety and health, and not citations. If a lockout problem or a chemical labeling inadequacy was discovered, the OSHA inspector would advise the company how to put in an effective program. This would make the inspector part of the abatement process. In this change, no citation would be issued, only a letter requiring the business to follow it up with a written program and confirmation of worker training. The inspection numbers would stay the same. The citation numbers would be cut by 95 percent. What would be the impact of these decreased citations when questions of OSHA effectiveness or funding come up? In our traditional way of thinking, citations and penalties means that OSHA is combating the enemy, industry, and conquering worker safety and health issues. This just isn't true. There is in this example, in fact, a negative effect, not a positive effect on worker safety and health.

From an industrial perspective, the OSHA regulations are confusing, overly demanding and detailed, impossible to comply with, and unnecessary. From this perspective, if an inspection can be kept on the "surface," the more complex and more difficult problems can be left undiscovered. Minor violations found on the "surface" are easily dealt with and abated, and cost little in interruption, dedication of resources and other related costs, such as attorneys. It is a clean, efficient way of doing business. Findings in the deep areas of compliance are just the opposite.

Besides, traditional thinking would argue that the resources thrown at compliance would be of no value to the business.

The Recordkeeping Mess

Injury and Illness Records. What was the purpose of keeping injury and illness records in the first place? It came from a need to establish industrial injury and illness comparisons. If your business is in the nonferrous metal industry, what injury rate is average, below, or above average? Why was this important? It's like competing in the global economy. If you have nothing to measure the quality or price of your product against, you have no idea if you are doing a good job or not. Through comparison, we know if we are average, better than average, a front-runner or need to improve and catch up. It's a very important piece of information for any business.

This injury and illness rate could also be used to provide trend data for a company. Are the statistics improving, staying the same or declining? As a company strives to improve any measurable indicator, trends are critical to knowing if they are improving or not. So, the original intended purpose of collecting injury and illness data was to help industry be successful by comparing industries to each other and trending the data over time.

What is the use of injury and illness data today? It is used to identify industries and businesses that need additional compliance efforts from OSHA. What data does OSHA use to identify target industries? What is a primary source of data for business focus programs such as the Maine 200 program? What do regional or state OSHA offices use to identify the businesses to be inspected and frequencies of those inspections? What is the common index displayed by those in support or opposed to OSHA to gauge success or failure of OSHA? Has the current use of industrial injury and illness data strayed from the original intent and purpose?

Recordkeeping as a Compliance Tool. As with most micro-management programs, recordkeeping has become bigger than life.

Instead of being a constructive measurement tool, it has become a destructive and manipulated tool that is shrouded with suspicion and distrust. This is sad.

Because of the use of recordkeeping as a compliance focusing tool, industry has become very concerned, even paranoid about it. The problems with this paranoia are that 1) the rates become more important than they are, 2) control of uncontrollable factors that influence rates are actively sought, and 3) those keeping the records tend to play terrible games with the data.

1) If your injury rate goes up, does it mean that your safety program is slipping? Heavens no! Rates follow Gaussian distribution; they have up times and down times that average at a mean level. This is a mathematical fact called deviation. If, due to normal Gaussian distribution, the rate increases, industry tends to go into a reactive mode to reverse the normal trend. Safety professionals know this management mode all too well.

2) During the late 1980's and early 1990's, did we experience an epidemic of cumulative trauma disorders in this country? The data would tend to support this belief and OSHA's push to promulgate an ergonomics standard in response would also tend to make you think that it had. If this is true, what sudden industrial exposure caused this epidemic? Was it the way we cut meat in the meat processing industry? Did that change? No, we've cut meat that way for a long time. Was it the expanded use of computer keyboards? No, they didn't suddenly explode on the market. What was it then? It was our knowledge of the disorder, our heightened awareness of treatment options and an explosion in the practice of elective surgery to resolve carpal tunnel syndrome. To a large extent, it was economic. It was a quick $2,000 outpatient surgical procedure that we now find doesn't solve the medical problem. In other words, this explosion in rates and the results, such as the proposed OSHA ergonomic standard were driven not by new or unregulated hazards in the workplace but by increased knowledge and economics.

How do you control the uncontrollable? You can't. But with the visibility and current uses of these rates, business actively tries.

3) Probably the most destructive effect is the game that tends to be played with the records. There is an odd practice in management. It's called "shooting the messenger." When bad news comes in the door, the messenger who brings the news is shot. You could argue that ignorance is bliss or that attacking the messenger is an easier and more convenient target. In reality, it's only a frustration- and ignorance-based reaction. After all, we are good at reactive management in this country. If you are the keeper of the OSHA log and you exist in a reactive management environment, which is very common, there is too much temptation to "cook the books." Call it rationalizing or finding any reason to not place a case on the log; it denies reality. If it belongs on the log, it needs to be there, regardless of management reaction, regardless of the current use of the data by OSHA, regardless of industry's paranoic sensitivity to the data. If it's a duck and quacks, call it what it is.

I talked about this problem at some length in my book *Total Quality for Safety and Health Professionals* so I won't repeat most of it here. It tends to be a problem with the way we manage our businesses in America and the misguided way we align the responsibilities for worker safety and health. It is solvable but it begins with knowledge, and then proper application, of effective management techniques.

Chuck Coonradt, in his book *The Game of Work*, established some basic rules for creating an effective and empowered workplace. In other words, provide a motivating and successful environment. One of his rules has been violated repeatedly in OSHA's quest for stopping alleged under-reporting. Coonradt says, if you want to create this kind of progressive environment, "don't change the rules in the middle of the game." Not only does it de-motivate those that are involved in the "game," but it creates a meaningless amalgam of data that is not comparable. This continual changing of the rules, of course, is a product of OSHA's not trusting industry to provide good data that will support their current use of the data, that is, as a means of targeting businesses to beat up on. Does this line of thought sound sick? It's one of those problems with the system again.

TODAY'S PERSPECTIVES

Political Perspectives

What are the current perspectives about what the OSHA Act created years ago? Is it still providing (or has it ever provided) value to worker safety and health? Are the current make up, direction, and administrative aspects effective for advancing worker safety and health in today's business world? The opinions differ. Looking at Congress may be helpful to us. After all, Congress only reacts to something that has high public concern or is mandated for review or reauthorization by statute. Without a statutory requirement to review and reinvent the Act, congressional actions are a good indication of overall perspective. Undoubtedly you are aware that OSHA reform has been getting a lot of attention in Congress over the past few years. The directions of that reform have tended to parallel party lines. Democrats want a bigger and tougher OSHA with more regulations, greater penalties including criminal actions, and easier promulgation of new or updated regulations. The belief that powers this direction says that industry is unsafe and seeks to sidestep worker safety and health in favor of profits. Republicans want a smaller and toothless OSHA that "consults" and teaches. The belief that directs this effort says that OSHA is a misdirected bully and that industry is a poor, ignorant, helpless pawn in the battle for worker safety and health. But, as with most partisan issues, reality and truth usually lie somewhere in the middle. That's why we don't have a one-party system. Our political system has the ability to provide perspective. However, having the power aspect thrown in, it tends to work like a pendulum.

Business Perspectives

What is the perspective of business toward OSHA? Their perspective says several things. First, the present system doesn't work. It is combative and gets in the way of important changes in business that must happen to ensure our economic viability in the future. Second, industry is not the bad guy of worker safety and health. No matter how widespread that opinion is, it isn't. Just because someone describes an animal as big and hairy that doesn't mean that it's a bear. It could be muskox or a Guernsey cow. Third, labor uses OSHA to beat up on industry and management. It's too easy to pick up the phone or drop a letter in the mail and call out the gestapo. This causes unnecessary work interruption and goes around the normal workplace resolution processes. Fourth, OSHA doesn't know its own regulations. They cite what is "in" at the time and they aren't consistent. Fifth, citations are merely a game for OSHA. Viewed like a traffic cop that has to write so many tickets per week, OSHA's need to find violations is critical to their purpose. Sixth, OSHA is inefficient and wastes money. Measured by industrial productivity and quality standards, OSHA would be out of business. It only serves as another government free-space where thousands of employees wait for publicly-funded retirements. Industry's perspective of OSHA isn't pretty.

Labor Perspectives

What's labor's perspective concerning OSHA? Are they a winner with the segment that so actively supported OSHA's creation? Labor's perspective of OSHA is mostly one of frustration. It hasn't been as effective at helping safety and health as it originally hoped. The regulation promulgation process is too slow. Congress has placed too many obstacles in OSHA's path, including cost-benefit analysis and OMB review. There are limited mechanisms for using current research information for protecting workers. There aren't enough inspectors, not enough inspections, and citations don't go deep enough. They only target easy and minor problems. The current situation and direction of OSHA doesn't have a chance of working. Not a pretty perspective, either.

Public Perspectives

What's the public's perspective of OSHA? Well, it's difficult to strip their opinions about OSHA away from an overwhelming frustration about the ineffectiveness and costs of government in general. Knowledge of OSHA in the general public comes mostly from media accounts that are sensational and one-sided. So, in most cases, OSHA is just one of the bags that the public has thrown up on the need-to-change wagon for Congress to deal with. No one can argue that there is a lot of public attention focused on changing the behemoth called government.

PROBLEMS WITH THE WAY IT IS

The problems are many—so many, in fact, that we could spend the rest of this book discussing them. That, however, is not our purpose. We've already mentioned some of the problems. So, at this point, let's talk about some of the major ones.

Regulators and the Regulatory Process Have Lost Touch with Reality and Impact

Recently at a national professional association conference, the head of OSHA, Joe Dear, was answering questions from the floor. Lamenting the considerable efforts required for compliance with the Bloodborne Pathogens standard, one attendee questioned if OSHA had any idea of the problems in protecting janitors to comply with the standards. Joe Dear confessed that before that question, he was unaware that the standard had any application to janitors.

Professional regulators just write language for industry to follow. They have lost touch with what the regulations they write involve, apply to, or cost to individual companies to comply with.

One of the most cited standards, the hazard communication standard is also a good example. What kills more workers in America—falls from above, ineffective or bypassed lockout protection, or not reading warning labels on chemical products found in the workplace? Silly question, right? If citations are supposed to be the main tool OSHA uses to protect workers, why do the majority of the "ten most commonly cited violations" come from the hazard communication standard and not from those that are killing the most workers? Answer: hazard communication compliance is difficult and cumbersome, and violations are easy to find.

There Is a Disconnection between the Professional Regulator and the Appointed Administration of OSHA

The appointed administration comes and goes. Those who have become members of the bureaucracy in OSHA stay and stay. There is a "wait till the next administration" belief in OSHA if the current verbiage from the appointed administration does not agree with ongoing agency efforts. Call it not listening or stonewalling; it happens all the time. This is ineffective and inefficient. It's not to say that each time the administration changes, the old agency directions should stop and new ones start. That would be even more inefficient that now. Nothing would get accomplished, especially in instances where the administration changes horses in midstream, such as the stance on the VPP in the Dear administration.

There Is Limited or No Applied Experience in Rulemaking

OSHA is comprised of professional, long-term government employees. The federal employment system works that way. With OSHA now well more than twenty years old, the length of service of many OSHA employees is nearly the same. Consequently, those who write the proposed rules have no experience in actually complying with them. Gathering this practical experience is the purpose of public comment. But, at that point, with one side liking and another disliking, it is too late. Actual workplace experience, long missing inside OSHA, needs to be part of the rule writing process.

The Continuity of Effort Suffers

Because of the odd mix between the bureaucratic bulk of OSHA and the changing administration, what is being said and done often conflict. Standards and programs pushed by one administration are left floundering for want of resources; new directions that can be opposite begin overnight. Other long-term efforts just sink into the mud and await a new administration. In other words, things get interrupted, changed, mired, confused, and directions changed.

A Failure of Logic

There is a basic belief that has saddled OSHA since its inception. This basic belief says that increased regulations and funding will equal decreased workplace injuries, illnesses and deaths. I challenge anyone to produce the evidence of this from the past twenty-plus years. The truth is, it cannot be done. This belief is not true. It is a natural outcome of our traditional way of thinking, however. Once a direction is chosen, wisely or not, more money thrown in that direction should make it more effective. We follow that way of thinking in almost every aspect of government, including public medical care, housing, the space program, the military establishment, and OSHA. Consequently, there is always a rush to the bank at appropriations time to keep the turf and enlarge the power base. In reality, our way of thinking is flawed. Bigger does not equal better. More money dedicated to any program does not equal greater effectiveness.

WHAT CAN WE DO?

OSHA

After working in industry and with management for many years, I have always agreed with the thought that making safety and health requirements a matter of law was the only way to make things happen.

After all, if it isn't enforceable and stated in a "thou shalt" format, management just won't take it seriously. It leaves too much up for interpretation and "what if we don't" arguments when it is not law. This is a common belief, a paradigm, that is prevalent in the safety and health professions and in labor. But one of the hardest challenges for us in our quest to improve worker safety and health is to go to the edge of our paradigms and dare to think differently.

An Outdated Regulatory System

Few would argue that our present regulatory system doesn't work; at best, it is terribly ineffective. I would add that there is no way that it can be anything else. Let's ask a simple question, one that has plagued us for many years. If our present safety and health laws are outdated, inadequate, and our mechanism for making or keeping them current is overly cumbersome and can't keep up with technological or workplace changes, why do we still have them? This question is almost heresy to our traditional way of thinking. But does it make any sense to say that a "thou shalt" approach is the only way to force safety and health into the workplace if our present laws are inadequate and have no way of catching up? Think about that for a moment. It's like having and enforcing a 50-mph speed limit in school zones where kids are being killed and saying that it is better than having nothing. This thinking is madness!

The Example of Superfund

In the late 1980's the Community Right-To-Know statutes went into effect (Superfund Amendments and Reauthorization Act or SARA). It was not a popular law to industry because it required them to expose their "dirty laundry," their chemical information, to the public. Now that's not saying that the SARA regulations are perfect. They are far from that. But, look at the positive impacts this required reporting has had. First, it has made industry more aware (at the highest management levels) of the inventories and releases of hazardous chemicals at their facilities. Second, it has made the public more aware of the potential harm around them and made them more involved in the community safety process. And third, the amounts of chemicals reported by businesses have continually decreased from year to year. Is community knowledge of chemical hazards and amounts important? Of course it is. It has been critical to the success of SARA. Has this approach been successful at making management more involved in chemical reduction? No one can argue its positive impact.

How is the approach taken by SARA different from the one taken by the OSH Act? It is self-disclosing instead of requiring some inspector to discover it or some event such as an injury or a complaint to point it out. There are public visibility and trial instead of one-on-one discussion and resolution. SARA rules themselves are a fraction of what OSHA has in print. There is little enforcement staff compared to thousands for OSHA and the state OSH programs. SARA has produced significant, measurable, and documented improvement, which is more than we can say for OSHA. These are significant and important differences.

Using the Same Approach

Why can't this same approach be taken for worker safety and health? Why can't we expand the VPP application to include and detail the important areas of worker safety and health, and require each facility to make a report to OSHA every so often? Why can't we just pitch the volumes of OSHA "thou shalt" regulations out and replace them with the simple General Duty Clause, reporting requirements, inspection language, and a list of accepted guidelines like ANSI standards for use as the "measuring stick?" Why can't OSHA publish an annual report on the state of worker safety and health at each facility, and give it to the public? Why can't we tie an OSHA facility ranking to insurance rates and create a list of preferred workplaces for businesses and potential employees? Why can't we think differently and take a totally new approach to worker safety and health?

What are the advantages of such an approach? The reporting alone would force dedication of resources and increase management knowledge of the status of worker safety and health in their facility. Having public visibility and OSHA ranking would create an incentive for management to commit resources and demand improvement. Self-disclosure would totally remove the traditional practice of requiring a hazard to be "found" before being forced to correct it. Using an expanded, modified VPP application for reporting, which already requires labor involvement, would ensure worker issues being addressed and their knowledge and approval of the process. This would force communication between management and labor on safety and health issues in all facilities, many of which today have little or none. Instead of OSHA dedicating so many resources and energy "reinventing the wheel" (writing, promulgating, getting approvals, etc., for regulations that cover existing guidelines), referencing acceptable guidelines and organizations reduces waste, assures a much more current measuring stick, and removes the unavoidable confrontation of rulemaking.

Periodic facility inspections to confirm reported information or gauge management and worker commitment, responding to employee complaints and concerns, and investigating reportable injuries and fatalities would need to be continued. This effort, however, would require a much smaller OSHA workforce to accomplish.

A New OSHA

If we created this new OSHA, what would we do with all the current OSHA and state OSH staff? Well, for the most part, they can be employed by industry, creating and updating the required facility reporting. In other words, they can become parts of the solution, not just those who point out problems. OSHA would have sufficient staff to change the focus from enforcement to teaching management and business how to improve worker safety and health. This is a critical and missing element in today's business world.

This approach is not OSHA reform. This approach is starting over. But, starting over requires us to admit that the present way doesn't work and to have the courage to begin to think differently. Don't take the word courage lightly. We have a great deal invested in the present system. This investment goes far beyond commitment of resources or momentum. It is ego involvement, a deeply personal investment. It will take *great* courage, perhaps greater courage than we have ever found in the past, to begin to think differently about our regulatory environment.

Here's another thought, seeing how we are talking about thinking differently: maybe the new OSHA would fit better in Commerce (if Congress doesn't kill that department) than in Labor? Think about it.

NIOSH

Let's make a bold statement. NIOSH is an inefficient and ineffective agency. From both a taxpayers and workers' safety and health perspective, it screams for a need to change. Understand that the original purpose of NIOSH was to provide research and educate professionals. The approaches taken to accomplish these two charges, however, have always been very different. Research has been accomplished in-house with NIOSH employees, NIOSH facilities, and NIOSH resources. Education of professionals has been contracted out to various universities. Although the university programs are far from efficiently run and the quality at different universities varies widely, this is and continues to be a significant need. We simply cannot ignore the connection between well trained safety and health practitioners and improving worker safety and health. Sure, these programs can be made more efficient and effective, but to have NIOSH which is even more ineffective and inefficient, managing this transformation and efficiency effort is a ludicrous thought.

What should we do with the in-house NIOSH organization? It should go the way of other wasteful, bad ideas. It should go into the garbage. If this same level of inefficiency, lack of productivity and poor quality were found in industry, it would be excised like a malignant tumor immediately. Why should we as taxpayers expect less? NIOSH is not the only source of research data. Actually, because of its inefficiency and lack of performance it isn't even a major player in the game anymore. Let the research be done by those who can efficiently do it. Perhaps grant money is the best solution, but continuation of the present NIOSH research and administration is crazy.

PUTTING OPTIONS, PRIORITIES, AND EXPECTATIONS IN ORDER

We've all been duped by our own paradigms. We have never questioned our options. Given "OSHA," we have all taken our turns pointing out problems with the system. Anyone can point out problems, even a three-year-old. Solutions, however, are much more challenging. Here again, however, our paradigms have gotten in our way. Blinded to all our options, we have traditionally chosen to "reform" OSHA, literally meaning "to form the same object again." This approach cannot work. It's too emotional to think differently though. We are too invested in the current OSHA and regulatory belief, and it is certainly too political.

Congress has taken another traditional approach. Because changing the system is too political and emotional (and they really don't know how to fix it), they choose to deal with it as an economic issue. It is like what management does in the workplace, making cuts in non-value-adding areas like safety and health. If this approach was one of the reasons the OSH Act was created in the first place and is one of the greatest frustrations to safety and health practitioners and labor, why do we consider this meat-cleaver approach acceptable for Congress?

This is a call to think differently. But, thinking differently about the regulatory environment is merely a subset to our need to think differently about every part of the system for worker safety and health. It always begins with basic questions and history. It always ends with questions that take us to the edge of what we know or can think. This vantage point is no less critical in our quest for an effective regulatory approach to worker safety and health. But make no mistake about it, the regulatory stance that we take is critical to our success.

What should the priority be for creating an effective regulatory environment for worker safety and health? It's right up there, near the top. Because by creating a new OSHA, we can provide a great deal of improved resources and powerful forces to bring about other necessary changes such as management and labor commitment.

Our expectation, however, is the easiest part. Our present expectation of having an effective regulatory environment need not change. In the present system, this has been an extremely frustrating expectation, one that is branded as idealistic. If we choose to let go of our expectations due to what we define as reality, we continue our frustration, and live with an ineffective system. If, however, we demand that our expectations, our dreams, become real, we release the power of positive change and begin to think differently.

SUMMARY

Current problems—what we should stop doing:

- Believing that a strong OSHA is the only way to improve worker safety and health.
- Condoning a grossly inefficient and ineffective system as the best we can expect.
- Reinventing the same regulatory process.

Fixing the system—what we should be doing:

- Questioning everything about our regulatory process, including validity, structure, and application.
- Questioning if the research portion of NIOSH is the best option today.
- Inventing new ways of creating visibility and recognition of worker safety and health excellence.

6

THE LEGAL VERSUS LIABILITY GAME

Ours is a society of laws. It has been argued that this is the one major element that separates us from barbarians. Barbarians have few laws. Civilized societies have many. If the number of laws is a measure of the height of any civilization, we should be nearing the highest possible level any society has ever achieved in man's history. More laws are added daily. Seldom are they removed. Looking at an example of one of the more memorable ones to emphasize a point, let's consider one that bans spitting on the sidewalk. Today the purpose is unclear and, as a result, the enforcement is never pursued. But if you know the history of this no-spitting law, it becomes a lot more clear. It was passed when men chewed tobacco in public and regularly spit the associated juices on the sidewalks where everyone walked. In those days, women wore dresses that were so long, they dragged on the ground. Knowing the history, one can see very graphically why the law was written. The need for this law today is, of course, not as obvious.

ARE THERE CORRELATIONS IN OUR SOCIETY?

Why aren't outdated laws removed? I believe it is because we truly think that the advancement of a society can be determined by the number of laws they have. Why else would we not routinely get rid of laws that just don't make sense anymore? There was an idea that surfaced a few years ago that stated that you couldn't write a law unless you got rid of one. It didn't receive more than a curious and humorous response. No one really took the idea seriously. We rationalize that as our society became more complex, it required more laws.

It's interesting that when our forefathers had the rebellious idea of starting this country two major things really angered them. They were aggravated enough to take on treasonous activities against Great Britain. One problem was the issue of taxation without representation. Who could blame them for getting angry about the number taxes placed on them, and the increasing cost of those taxes? The other was the number and layers of laws that complicated their lives. They had the complex laws of England that often were a mismatch to a developing group of colonies. Further, they had colonial laws that were enacted by the British hierarchy on the local level. There were just too many of them and it seemed that the colonists had no say in their creation. Who wouldn't be upset if they were in a situation like this?

Are *you* upset? Look at the example that was just used and transpose it to our lives today. Kind of fits, doesn't it? Why aren't we outraged like our forefathers were? Remember the story about the frog? Dropped into hot water, the frog will immediately hop out. However, take the same frog and heat the water slowly and it stays and is cooked. The laws have come so gradually and continually, that we, like that frog, have become desensitized to our situation. Obviously, we are a whole lot more lethargic than our forefathers. It is probably due to our easy lives compared with their difficult ones. If we have our TV remote and a bag of chips to eat on our sofa, we accept all the garbage around us. We just tune it out.

LAW: PURPOSE AND CORRELATIONS

What is the purpose of law? Let's get away from the legal jargon definition because it comes from biased origins—lawyers and law makers. Simply stated, law is there to tell us what is acceptable and not acceptable to society. It standardizes our behavior so that we can all live together. Look at the purpose of laws like the rules of a game such as football.

What kind of havoc would it bring if one team showed up at a game with big spikes on the tops of their helmets? How about if they used a laser-guided mechanical catapult-type device to kick 90-yard field goals? Let's say that one team demanded that they be awarded 10 points for each touchdown? In this "anything goes" game, it wouldn't resemble what has become a part of our American society. It would more closely resemble the Christians and the gladiators. Laws standardize our behavior so that we can live together.

Because of this, laws have always had a high correlation with our behavior in society. In other words, most of us, a vast majority in fact, follow them. Why? We follow them because they are enforced. Some portion of our government which manages the society is designated to enforce the laws. Because of that, laws are also linked to some sort of penalty if they are not followed. Breaking some laws gets you a ticket and a civil penalty. Some can get you put into jail. Others can bring death as the judgement of our society. The more severe the consequences of a law being violated in that society, the greater the penalty or punishment. This doesn't always hold true, but generally it's a good correlation.

Knowing this, there are two other important correlations. Those laws that have less severe penalties are broken more often than those that bear severe penalties. Also, those laws with lesser penalties receive fewer resources dedicated to their enforcement than do the heavy hitters. Look at a comparison between jaywalking and murder. How does the punishment of infractions of these two laws compare? Jaywalking is punishable by a civil fine of $25 or less. Murder, however, is a capital offense. Is there a correlation between the number of times each is broken and the punishment? Walk around any downtown area. Jaywalking can be observed regularly on almost every block. But every time someone is murdered or a body is found, it is big news in the newspaper. Can one observe a murder or two on every downtown block? No. There is a wide difference between the numbers of infractions of these two laws. Is there a correlation to the government's dedication of enforcement resources? What are the odds of receiving a ticket if you jaywalk? The odds have got to be less than one percent. What are the odds of getting caught if you murder someone? Much higher because there are so many resources, including police officers, crime laboratory personnel, and tests (even specialty tests like DNA tests) thrown at solving the crime.

Society looks at laws as black or white issues. Either you comply with the law or you break it. You cannot be partially in violation of a law. It's like the impossibility of being semi-pregnant. Either you are or you aren't This is why we have judges and juries to resolve these issues. If there is no "reasonable doubt" established, you're guilty. It's a black and white issue.

LIABILITY IS DIFFERENT FROM LAW

Liability is much different from the laws of society. Liability is more subjective, more variable, and has no true tie to a penalty. It isn't generally written in "thou shalt" wording. It just isn't as clean-cut as law. Why? Because we don't pass liability issues like we do laws and, generally, liability issues are resolved in a tort action, not in a criminal court. Liabilities are much different from laws.

Where laws are "thou shalt" issues, liabilities can be said to be "thou should" issues. Because laws are "thou shalt" issues, they have designated penalties should you choose to violate them. Liabilities are more an issue of common thought or the practice of common law. It's a gentleman's sport. If you should have and didn't or shouldn't have and did, this is a liability issue. It's like the advantage rules of soccer. If a player gains an unfair advantage on an opponent by violating one of the advantage rules, the game is stopped and, as much as possible, the opponent is given an advantage that cancels the one taken. If an advantage rule is violated and no one gains an advantage, the game continues without interruption.

Liabilities normally only have civil or monetary penalties. For that reason we usually have insurance policies to protect us if we should cause an automobile accident, have a guest slip and fall on our door step, or if our dog bites the neighbor. Seldom do violations of liability issues result in criminal penalties unless gross negligence is involved and one can reach into the criminal statutes (laws) for punishment.

There is another significant difference between liabilities and law. In law, you are guilty or innocent. There is no gray area. In liabilities, there is nothing but shades of gray. If you're not innocent, it is *always* s matter of degree. The tort system is set up to determine innocence or the degree of guilt. Correlating with that degree of guilt, theoretically, is the amount of the award. The less guilty, the smaller the award, and vice versa. Liability is not a black or white issue. It is white and a whole lot of varying shades of gray to almost black.

There may be some interpretation required by law like: was the killing premeditated, an act of rage or in self-defense? These are only used to determine whether the law was broken or not. Liabilities, however, have a much greater burden of proof. Two significant issues pop up—knowledge and negligence. Did the person have knowledge that this could have had a detrimental impact or not? Knowledge is an important bridge to any consideration of negligence. Obviously, if the person was uninformed of the hazard, how could they be negligent? With knowledge, however, comes the issue of proving negligence. In other words, he knew better but stubbornly refused to do anything about it. Knowledge and negligence issues offer more challenges to liability claims.

HOW DOES THIS APPLY TO WORKER SAFETY AND HEALTH?

Why did we go through this primer on laws and liabilities in a book on worker safety and health? Prior to the OSH Act, worker safety and health was a liability issue to business. Technically, there were no "thou shalt" regulations associated with it, only "thou should" issues as a means of controlling monetary risks. As we do today, they had insurance policies called workers' compensation. With the advent of the OSHA regulations, industry had the specifics of previous liability issues consolidated into the details of law. Along with this change came the aspects of guilt or innocence, black or white, that accompanies laws and not liabilities. With the inception of OSHA, it was a very different world for business.

This guilt or innocence, black or white instead of the negotiable gray was one of the major points argued by those proponents of the OSHA Act. Liabilities weren't enough of a force and there was too much gray to improve worker safety and health. Opponents feared that the weight of having the "thou should" aspects turn to "thou shalt" regulations would become an overwhelming burden on business. What they really feared was trading all that gray area for only black.

This concern of business was warranted when guidelines or recommendations that bore little or no liability risk became law and, thereby, had the power of enforcement. Good examples would be split toilet seats and handrail height. Personally, I am unaware of any liability torts filed against employers because they did not have split toilet seats. Factually, the risk just isn't there. Consequently, it was common for business to just "blow off" silly regulations like the split toilet requirement. Likewise, what is sacred about 42 inches height for hand rails? I'm sure that some analyst could make some mathematical case using anthropometric data and balance ratios of the human body to say that 42 inches is a good height for handrails given 90 percent of the population. But wouldn't 44 inches be safer by the same mathematical model? I also doubt if 40-inch handrail height would significantly raise the risk to workers. The point in these two examples is that as recommendations or guidelines (which, incidentally, they were written to be), they establish the lower or most near-white segment of the gray area. It is far from black. Little liability risk exists. Using or not using these guidelines becomes a decision of business, based solely on need and employee concerns, balanced against the cost and the financial wellbeing of the business. Made a matter of law and cited by OSHA, however, as both of these examples have been many times (until the split toilet seat rule was rejected), compliance is black and white. It is no longer a business or employee interest decision. It became a "thou shalt" compliance issue. Multiplying the number of handrails and toilet seats in business, it wasted a lot of money that could have been spent correcting real hazards.

SAFETY AND HEALTH LAWS VERSUS LIABILITIES

In our society, it seems that one law is never enough. For example, common sense tells us that it is enough to say "don't steal." If this is the law and violations of the law are black or white issues, our court system is more than adequate to determine guilt and apply an appropriate penalty. Not so. The statutes that encompass and define the details of "don't steal" requires volumes. And, the details of law continue to build daily. But many argue that it is too easy to pass additional laws and add detail to issues such as stealing and murder because they are emotional issues, considered to be blemishes on our society. We are too ready to pass new laws and detail to attack these emotional issues.

When the promulgation aspects of OSHA were being written, this was a significant issue. Keep in mind that OSHA regulations do not carry the same emotional weight as do theft and murder. There were a lot of "checks and balances" put into the law in an effort to make sure that the OSHA regulations made sense, would not overly burden industry and would have the desired outcome. What we ended up with (including a lot of extra baggage of OMB review and cost-benefit analysis) was a fairly inefficient and overly laborious promulgation process. Nevertheless, OSHA regulations continue to stack up, defining more and more "thou shalt" details for business to comply with.

Liability issues continue to impact business too. Whereas the legal requirements, the matter of law, are dependent upon being written or referred to in the regulations, liability issues are different. Remember that liability usually requires two aspects—knowledge and negligence. Obviously, the knowledge issue is extremely important. What is it dependent upon? In safety and health risks, knowledge is dependent upon current guidelines, research findings and public fears.

We have a well-developed guideline development system for safety and health issues. Recommended guidelines are developed by many sources, including the National Institute for Occupational Safety and Health (NIOSH), the American National Standards Institute (ANSI), the American Conference of Governmental Industrial Hygienists (ACGIH), the Chlorine Institute, the American Gas Association (AGA), the American Petroleum Institute (API), the American Society of Mechanical Engineers (ASME), Underwriters Laboratory (UL), and the American Welding Society, to name a few. In fact, ANSI lists 108 guideline-writing organizations that they use.

There is a loose association between a published guideline and business knowledge. It goes something like this, "If it's published and available, you should have known." In most safety and health liability torts, ignorance is no defense. Business should have known.

There are also a lot of research findings out there. Created almost daily, new studies are published on correlations between various exposures and animal or human health effects. There are two troubling problems with current research. First, a lot of studies contradict the findings of other studies. And second, how strong a correlation is necessary to make a "reasonable" business knowledgeable and, thereby, act on that knowledge?

The most powerful study is, naturally, an epidemiological study because people are used. However, epidemiological studies that use large populations and encompass long time frames are rare. Because of that a lot of animal testing is used where mega-exposures over a short time period are used to simulate low exposures over long periods. This is where a lot of the contradictory findings are rooted. This is an important difference between research and published guidelines.

Public fears or concerns are even looser, but much more volatile. Ask any business that has been "tried" on a public fear issue in the media. This is serious business and carries significant liability risks. In short, the rule of thumb is: "If the public is concerned about something, the business had better be also." Knowledge of the concern is enough.

BOTTOM-LINE ISSUES TO BUSINESS

A Risk Equation

In the business world, the difference between legal requirements and liabilities often becomes insignificant. Business' interest becomes a risk equation on the balance sheet. What is the potential impact to the bottom-line? Obviously, there are significant risks or ramifications of noncompliance with regulations and not performing in a socially responsible mode. The ramifications of noncompliance are mostly civil. Monetary penalties have been the common result. Lately, however, these penalties have become significantly higher. Looking at the multimillion dollar penalties issued against large corporations for recordkeeping violations or use of the egregious multiplier when calculating penalties gets the attention of business. Realistically, penalties in the hundreds or even thousands of dollars have minimal impact on bigger businesses. OSHA knew that when they proposed the egregious multiplier. The thought was to get the penalties up into a region that got the attention of business. In reality though, it just deepened the trenches, increased the antagonism, and employed more attorneys.

Criminal Penalties

Another avenue was attempted, but failed. Borrowing from some of the applications of environmental law that allow for criminal actions, in 1994, a consolidated effort was sponsored by Senators Kennedy and Metzenbaum to add criminal penalties to OSHA reform legislation. With the dominant party change in Congress the next year, this legislation went straight to the back burner and kinder OSHA reform legislation appeared. This being the case, however, there are some important lessons for us. First, criminal penalties already exist on the criminal side of law. If a supervisor or corporate CEO is responsible for the death of a worker, these criminal statutes can be applied. For sure, application and proof are not as easy as if they were a part of OSHA law, or the CEO was holding the "smoking gun," but the laws already exist. This realization goes back to the argument: if we already have a system in place, why does it make any sense to add details and regulations that say the same thing within a specialized system? Recognize that it fits the way we traditionally think, but does it make any sense?

Second, the popularity of OSHA reform doesn't seem to be a single party issue. The only aspect that seems to be party-oriented is the strength of the regulatory language. The lesson here is that whatever regulatory language OSHA follows (for example, with criminal penalties) depends to a great extent on what political party or ideology is in power. From that lesson, we cannot become too complacent, thinking that criminal penalties are gone and will never threaten American business again. That belief is naive, knowing the current volatility of the political climate and the voters' seeming impatience.

RAMIFICATIONS OF LIABILITY ACTIONS

The ramifications of being caught in a liability situation are a whole lot more complex than being out of compliance with law. Whereas laws are linked to penalty provisions, liability torts are not. The obvious primary ramification is being ordered by a court to pay "through the nose" in a liability tort judgment. There are two areas of damage assessments in such a judgement—compensatory (direct damages) and punitive (indirect damages). Because of the magnitude of the difference between these two areas of damage awards, compensatory damage awards have become of secondary importance when compared to punitive damage awards. Too often in today's court system, these punitive damages are being used to get the defendant's attention or to makeup for all the alleged "others" who didn't initiate suit.

But the ramifications of liability actions can extend way beyond the monetary issue or make the monetary damage significantly greater. These other ramifications can be as damaging, if not more so, to an involved business. I have grouped these into five areas: costs of going to court; our present unfriendly court environment, risk decisions following actions;

negative public opinion; and state-by-state and region-by-region confounders.

Cost of Going to Court

Going to court is *not* cheap. It never has been. Defending a liability tort can be extremely expensive. This cost must be weighed against what the monetary risks might be should the business lose. Obviously, it is a poor business decision to spend $250,000 dollars defending against an action that might receive a $5,000 award. That is, unless there is a significant precedent-setting issue at stake for the business. Putting a Clarence Darrow-type of attorney team up against a local attorney may get a sympathetic verdict for the plaintiff regardless of the strength of the defense. Also, spending huge amounts of money on minimal risk tort action might also send the wrong message to the court system. After all, if the business is willing to invest that much money, maybe the historical awards have been too low.

The Court Environment

The court environment has changed in two significant ways. Both of these can notably affect a business if a liability issue surfaces. First, unlike the court environment prior to the existence of the workers' compensation system, there are plenty of attorneys available to take liability cases. It seems that at least two attorney ads appear each evening on television, regardless of which channel you watch. "If you have . . . call us today," the messages say. They also say, "If you don't win, you pay nothing." Not only has the availability of attorneys increased, so has the public knowledge of this availability. And, if it apparently comes at no risk to those initiating the suit "unless you win," the initial cost barriers are removed for would-be liability seekers. The second way the court environment has changed is in the "friendliness" to businesses. Would you say that the courts are more or less friendly to business today? Look at the amounts in punitive awards that have been recently given to plaintiffs. In

Texas, an owner of a foreign luxury car got a $4,000,000 award because the dealer touched up the paint on his new car and didn't tell him before he sold it to the plaintiff. How much did the woman get for spilling hot coffee from McDonald's in her lap?

These are merely two examples of a growing avalanche of high punitive awards given to plaintiffs. These excessive awards have additional effects that are negative to business. Excessive awards tend to create a mind-set in would-be jury members (the general public) that business is bad, conducting business with a blatant disregard for public safety and health. These awards also set an acceptability of excessive levels in juries' minds. "If they awarded $6,000,000 for that, then this has got to be worth at least $8,000,000."

Recently, some states have been trying to limit the amounts of an award that can be given. Texas is a good example. The constitutionality of this law will ultimately be decided in the United States Supreme Court. It should be interesting. Unfortunately, however, it is also an example of our traditional way of thinking in this country, passing more laws instead of fixing the problems.

Risk Decisions Following Actions

Following a liability action, a business tends to think a little differently. Odds tend to mean nothing at that point. Given that the odds of a liability action might be one in 3,000,000, having just stared down the barrel of a tort action gun, the odds become a lot closer to one in one. It tends to turn a company ultra-conservative. Conservativism is an ugly thing when a business gets caught up in it. It affects how the business spends money, where they place resources, the oversight or inspections that are required, and generally, how they look at the world. It changes their thinking from "opportunity" to "risk." It costs that business money and efficiency. Either can be the difference between future success and failure.

A mid-size manufacturer of electric switch gear was sued when one of their pieces of switching equipment exploded under a load and blinded a worker. Anyone who knows high voltage switch gear knows that there is a hazard when re-energizing the unit. The electrical surge has been known to blow doors off of equipment. It is common practice to mount operating handles on the sides of the units, not on the front of the box, for this reason. Electricians are trained to use their nondominant hand, stand to the side of the unit, and look away from the unit. This worker either didn't know the hazard or didn't follow proper precautions. The tort action was decided in favor of the worker. The reasoning was that the manufacturer did not place adequate warning signs outside the unit. As a result, did the manufacturer change the box configuration and the signs? Of course, and added 15 percent to their costs by doing so. This raised the cost of their units. They also sent warning letters out to everyone they had sold their units to. As a result, their market share dropped, due to the cost of the units, the warning letters, and the adverse publicity the manufacturer also received. Seven years later, this manufacturer filed for bankruptcy protection.

Public Opinion

When was the last time you saw a story about a business successfully defending a liability action on the 10 o'clock news? You haven't recently? Have you ever? Probably not, because the media loves liability actions, but only if the business loses. This one-sided media attention creates unavoidable negative public opinion about business in general. If you happen to be the unlucky business that gets to pay or appeal an award that gets news coverage, you have a severe public opinion problem. Look at excellent examples of this: Exxon with the *Exxon Valdez* spill; Union Carbide with the Bhopal, India, deaths and the subsequent Institute, West Virginia, release; Phillips Petroleum with the Texas explosion that killed many contractors. The list goes on. Factually, it takes a lot of time to come back from a case of negative public opinion. Sometimes you can never come back.

State and Regional Differences

How would any state-by-state, region-to-region differences be confounders to liability ramifications? It's a fact that some states have different political leanings. This can have a large impact on liability actions, court friendliness, and awards. This is why when a national liability case is filed, it is common to make the filing in a court system that leans the way you want them to. It's called "getting a leg up" on the opposition. In cases where they are state-specific, the state-by-state differences can also be significant. If the business is unlucky enough to have the liability case in a state that likes plaintiffs and dislikes business, that's unfortunate. However, if the business is lucky to have the tort action in a state that likes business, which is rare today, the business may have the advantage.

THE LINES BLUR

If this whole issue concerning legal requirements and liability risks to business isn't complex enough, there is one irrefutable reality emerging from today's world. The once clear separation between the "thou shalt" regulations and the "thou should" risks is narrowing. In many areas the gag has all but disappeared. This is good and bad—bad from the vantage point of costs and traditional management, good from simplification of the issues.

Liability Risks

Liability risks are more volatile and dynamic than legal requirements. Liability issues are more often than not, the higher standard of performance. Obviously, meeting the higher standard can require higher costs to business. But, looking from a total risk control strategy, it can produce a higher premium to business also. Knowing that the costs of dealing with a liability issue far outrun those of a regulatory compliance action, minimizing the risk of this magnitude of expense can be financially attractive to a business and free management from making decisions based on fear. But traditional management thinking can also be a roadblock to narrowing this gap. It's a paradigm issue. If traditional management beliefs dictate that liability issues are totally voluntary and that there will always exist a line between what you have to do and decide to do, the decision to move toward the higher standard can get lost in the shrubbery of tradition. This is a change issue also. Just as the practice of management must change to be successful in the future, traditional management paradigms must also change.

Positive Effects

The blurring of lines between legal requirements and liability issues can be a tremendously positive aspect to a business. Consider the analogy of a decision to buy a Chevy or a Lincoln. You have to have a car, but the choice to purchase the higher standard, the Lincoln, is totally a matter of choice. As market demands move the Chevy closer and closer to the quality level of the Lincoln, however, that decision becomes easier. Ideally, at the point that the difference between them disappears, you have one choice. As the gap between "have to" and "need to" becomes indistinguishable, the confusion and difficulty in dealing with two levels goes away. Management becomes simplified.

Consider some examples. Let's say you have a metal pickling process that uses hydrofluoric acid and nitric acid. Hydrofluoric acid falls into the "thou shalt" coverage of the Process Safety Management Standard (29CFR1910.119). Nitric acid at less than pure concentrations does not. Dealing only with the required actions under the standard, the process equipment that has hydrofluoric acid in it falls under the higher standard requiring Process Hazard Analysis (PHA), written procedures, trained workers, management of change programs, mechanical integrity programs, etc. The nitric acid requires none of these hurdles. Obviously, both acids carry a great deal of risk in that they are both dangerous chemicals that require very good operating practices and have similar hazards. Where is the line between the requirements imposed on hydrofluoric acid and the liability risks that concern nitric acid? They don't exist.

Compliance

In their compliance way of thinking, some OSHA inspectors have begun to recognize this blurring also. Although OSHA is unable to cite them under the chapter-and-verse of the 119 standard, chemical processes that have parallel hazards, but have control programs that are less than those required by Part 119 for covered chemicals, are being cited under the General Duty Clause.

It is hard to argue that these chemicals don't constitute a known hazard, especially if the facility cited also has complied with Part 119 requirements for other "listed" chemical processes. But, citations under the General Duty Clause always invite contestment, are decided by the technical knowledge or ignorance of an administrative law judge, and get significantly watered down along the long trail to resolution.

Looking at a previous example, the Hazard Communication Standard is another good example of the blurring lines between requirement and liability. What are the specific requirements concerning labeling of chemical containers?

In a performance standard, they are vague at best. How many ways can, or is, this requirement being met in industry? Millions of ways. Some are good, some are bad. Some look more like a Material Safety Data Sheet (MSDS) than a label. How is strict compliance with the standard's labeling requirements judged? Normally, it is determined by a complaint or by a mishap, such as an exposure or injury. What is the liability risk involved with poor labeling? Obviously, it is very high. How much difference really exists between what the law says you must have on a label and what risk avoidance dictates? Can you see *any* difference?

RISK-BASED DECISION CRITERIA

Let's bring our discussion down to the basics, the often ugly dollars-and-cents issues associated with risk. Too often management gets tagged with this level of micro-management thinking concerning regulatory and safety risks. Sometimes it's true. Sometimes it isn't. But when we discuss risk-based decision-making, we need to look at this most basic level of decision-making criteria. The question sounds something like this: "What is my financial risk if I don't choose to do it?" Obviously, there is another question that usually follows: "What is the probability associated with that risk?" But, probabilities are so difficult to calculate. They are more a function of luck, how well you get along with your employees and the competency of the OSHA inspector. So, we will focus on the first, more quantifiable question of financial risk.

The answer to this question must consider a number of aspects.

What Costs More?

What is the cost of compliance, machine guarding, modernization, facility improvement or modification, system or equipment upkeep, and penalties should noncompliance be cited? What is the cost of reducing the liability risk, fixed and ongoing, and the costs if a tort action is initiated including attorneys' fees, work interruptions, settlement or judgment costs, etc.? This is a financial question. From a balance-sheet perspective, a manager needs to at least ask these questions. Sure, there are some tradeoffs. It makes no business sense to spend millions correcting a regulatory issue or liability risk that is next to nothing. This, of course, is removed from the human side of the questioning process. This is totally economic.

What Is the Viability of the Business?

If a company is on the financial ropes and is struggling to make the payroll each month, this is greatly different from a business with a 20 percent return on investment. This is too obvious to spend much time discussing here.

In Which State Is the Business Operating?

There are important state-to-state differences. Some have an approved state OSH program; others are covered by federal OSHA. The penalty structures in different state plans are different. Some are significantly higher than others. OSHA staffing levels differ widely between state programs. For states covered by federal OSHA, the distance to the OSHA regional office and density of industry in your area are considerations. What is your state's or region's contestment settlement history? Is it liberal or conservative? Does it "tow the line" or is it forgiving? In state OSHA programs, what is the pay scale for inspectors? Some states have good pay levels and keep competent inspectors. Some have lower pay scales and high turnover that impede competency in worksite inspections. Does the state or region use specialized inspectors or are they cross-trained? There are many state or regional factors that can impact risk-based decision making.

What is the openness to liability tort actions in your state and what is the success rate? What is the average award level given in successful cases? Are liability actions rubber-stamped to require extended litigation efforts? These issues can have tremendous impact on whether liability risks are high to a business.

What Is the Speed of Resolving Legal or Liability Issues?

Compliance issues that receive citation and penalty are resolved fairly quickly when compared to settling liability torts. Complex compliance issues that are contested and appealed can take a lot of time. But, on the average, the speed of bringing compliance issues to resolution is much faster than liability issues. Additionally, compliance penalties are pretty well known. Liability torts are not as easy. Depending on many factors, settlements can range upwards from one dollar to even millions.

What Are the Legal Requirements for Subsystems such as Workers' Compensation and Liability?

Resolution of OSHA issues goes before an overloaded Review Commission. Time for resolution is also a consideration. State OSH and workers' compensation subsystems generally use the ALJ means of resolving issues. These can be more highly political and one-sided than OSHRC. Appeals of state ALJ decisions are handled by state court systems. But, generally, they can be appealed for as simple a reason as you don't like the decision. Liability torts, however, fall under the normal judicial process. Appeals are much more difficult and usually have to be made on some legal ground such as excessive award. Each system for resolution of problems is different. Each has advantages and disadvantages. In any case, in making risk-based decisions, these are considerations.

What Is the Availability of Attorneys to Handle Litigation?

The availability of attorneys for litigation is not like it was in the early days of the industrial revolution. Factually, with the number of attorneys who advertise, one could argue that there is not only an abundance, but there is a feeding-frenzy out there. This also ties in closely to the state-by-state differences in court openness, decisions, and award levels. Those states that have a real openness to such actions, a majority of decisions for the plaintiffs, and higher than average awards, tend to have more attorneys and the risk to the business is considerably higher.

What Is the Role of Cost Drivers?

Just as at the birth of the workers' compensation system, the cost drivers are a significant consideration. Remembering it was the cost drivers that brought about workers' compensation in the first place. But in today's world, those cost drivers are very different. With OSHA's use of egregious multipliers and ridiculously high penalties for differences in recordkeeping opinions, these can be significant concerns. Arguably, the probability of such a concern is not high. But if smacked with a big one, it can make or break a business or more realistically, a management group. Cost drivers in liability cases are also much different today. In the early 1900s the awards were much more in line with expected wage losses. Today, the courts want to "hurt" the plaintiff. No longer is it a common-sense reimbursement or costs issue, courts award mega-judgments to make an example of and hurt the plaintiff for being mean or worse, especially if you are considered to have "deep pockets." Cost driver risks for liability cases, therefore, can easily be in the millions of dollars. The costs of defense can be just as impressive.

When we stop and look at risk-based decision-making, it tends to add a new dimension to worker safety and health concerns. One begins to realize that the world we thought we created with the workers' compensation system, a "no-sue zone," today is a myth. More and more worker safety and health issues are being settled, not in the created workers' compensation system, the exclusive remedy, but in the tort courts. And this trend seems to be increasing in our litigious society in which we are compelled to blame someone else for everything bad that happens to us.

OTHER RISK-BASED CONSIDERATIONS

Liabilities are based on current information, as published in guidelines and literature, and legal requirements are a function of promulgated law. Comparing the two is an inequality. Promulgating or changing laws takes so much time and is so political that it causes most to say "why bother." On the other hand, guidelines seem to be published with relative ease and at regular intervals. Why is this a significant concern to business? Dealing with the risks of noncompliance is like keeping up with a turtle. Minimizing the risks associated with liability is more like racing the rabbit, which is a very fast rabbit. Being so fast, there is another consideration. There are more errors and overly-conservative estimates borne in guidelines. How does one deal with that? Well, just as liability issues are still a function of knowledge and common sense, so must be the measuring stick of what business considers a significant risk and what they do not. There must be a common-sense approach to accepting or rejecting information based on the degree of industry or professional acceptance.

The regulatory direction toward criminal actions for noncompliance is also a consideration. Real or not, the threat to upper management is high, depending of course, on the level of management paranoia.

PAST THE POINT OF NO RETURN

Having traveled this far in our analysis of the legal and liability risks associated with worker safety and health issues, there should be one obvious point. The once clear line between what you have to do because the law says so, and what guidelines and current research say you should do, has all but disappeared. Looking at today's risk issues and considerations, the playing field should be pretty well level for management. Then why is it that we in American business continue to wring our hands and protest the promulgation and change of regulations when the true risks to business are the same? Why do we waste time, energy, and expenses chasing a meaningless issue? We do so because of tradition, old-school management, and because worker safety and health is not a cultural value in American business. We've really never stopped to think about it and plan an appropriate strategy.

Tradition

Look at the stance business took in the debate over the OSH Act. Business has always taken the stance that any law published by OSHA is not good. It is a black and white issue. This traditional stance sets our business paradigm against OSHA. And American business is traditional. Even when shown new, different, and successful options, like total quality and OSHA partnership, we don't change. We rationalize that what has worked in the past will be successful for us in the future. It's a "head in the sand" approach to change. But it is traditionally American.

Old-School Management

America has the paradigm that only the companies that are driven by decisive, single-minded, autocratic leaders will be successful. It must be true, as common as these individuals are at the top of American businesses. So, everybody in the chain patterns themselves after what is considered successful. In some ways, there is a lot of natural selection involved in that only those that display this tendency tend to be promoted into management and up the management chain. In some companies, there are notable changes but mostly in selected sectors of the business. We tend to hold to the Drucker model of management as decisive and economically driven. Huggers and kissers just don't make it in American business.

These old-school management styles and beliefs place the white hat on business and the black hat on government. Additionally, old-school thinkers are stuck in the next-quarter economic mind-set. We don't seem to get into long-range thinking. Everything is dependent on the next quarter's financial figures. So, we don't analyze situations like worker safety and health risks farther out than three months. How can you come to grips with the blurring lines between legal requirements and liability risks if your frame of reference is a month or two out? You can't because you need to look far into history and far down the road to gain critical perspective, as the Japanese do.

Worker Safety and Health Is Not an American Cultural Value

This statement is not intended to insult anyone. It's a simple fact. We coined the phrase many years ago that said "Safety is number one." This is total "hog wash." Worker safety and health has never been even close to number one. The reality of our economic system alone should convince us that profit is number one. It has to be! The other business priorities can be placed in a predictable order behind profit. Production is number two, costs are number three, quality is number four, environment is number five, worker morale is number six and then come worker safety and health issues. Cold, yes, but it's true. This is a significant reason why we haven't taken the time to analyze the disappearing line between legal responsibilities and liability risks for worker safety and health. It just isn't a value or a priority in our American business culture.

SOLUTIONS, NOT BANDAIDS

Ever heard the saying, "When things start rolling downhill they just pick up speed"? Well, this statement is important for us when discussing the differences between legal requirements and liability risks. It's a simple fact that the incidence of liability tort cases in America for worker safety and health issues is increasing. The rate at which it is increasing doesn't matter. The trend line is going up. Our traditional "head in the sand" approach, not thinking or addressing this issue, will not change the trend. Ignoring any issue has seldom worked.

Once while driving a windy canyon road, my wife came around a curve and dead in her tracks was a herd of cattle. Not having the distance needed to stop and being very unsure of what movement the herd would take, she chose the only option she thought was available to her. She closed her eyes. Timidly opening them a second or two later, she found herself still continuing down the road with only cows in her rearview mirror. How did she do that? A thousand other drivers would have ended up with at least a cow or two in the front seat with them. It was just luck. Can we bet our stance concerning this disappearing line between these two issues on luck if our economic success as a nation is dependent upon the success of our business sector? It would be naive for us to do so, but it would be traditional.

America has another tradition. We tend to put bandaids on problems instead of making them go away permanently. We have never had the patience to study a problem for a long enough period of time to consider all the options and choose the best one. We look at the most expedient answer and the least costly. Because of this, too often we don't even get the problem right, much less the answer. So, we waste resources, time and money. The Japanese do things differently. Their term "to solve a problem so that it can never appear again," *poke yoke,* is very much a traditional way of doing business for them. They tend to be patient, a lot more patient than we are. But, if this is an important issue to our future success because the liability action trend is increasing and will only cost us more in the future, can we afford to take any other approach than to *poke yoke* it?

How do we find solutions then? First of all, we need to get our priorities straight. If worker safety and health can have significant impact on morale, environment (because it has to go through the worker level before it gets to the environment), quality, costs, productivity, and ultimately, on profits, how can we continue to treat it as a distant, or someone else's, priority?

Second, we need to step back so that we can view the entire forest. Too often we get so close that we can only see individual trees. We become myopic. We lose the important lessons that history can teach and we fail to look far enough ahead to plan a successful strategy. We only focus on today and the next balance sheet. This is not an easy tradition to change. But we must do so.

Third, once we recognize that the difference between what the law says we have to do and what we are at risk for only exists in our minds, we need to have the courage and commitment to act on our clear vision. It is too easy to fall back into the "why should I" economic-based myopic way of management. We can't lose our grip on what needs to be done.

ONLY IN OUR MINDS

Focusing our attention on complying with the regulations has always had the aura of security for business. If we are in compliance, we are doing okay. But safe in our compliance cage where OSHA cannot attack us, we forget to look around at what's inside the cage with us. We fail to see the crocodile of liability risk that lives in that same comfortable cage with us. If we are lucky, the crocodile won't be hungry. But as biologists know, crocodiles are extremely unpredictable. They tend to kill and eat even when they've just eaten. Like liability risks, crocodiles just seem to have a nasty streak.

The world where legal responsibilities were the measuring stick for worker safety and health issues is going fast, if it's not gone already. Just as American business is being forced to embrace the tougher standards of quality and costs, we need to embrace the tougher standard that protects us from the high and unpredictable risks associated with our liability. In fact, to control costs in tomorrow's business world (which we must do to be competitive), we must discard our belief that compliance is enough. Actually, it isn't even close. We can't afford the risk.

SUMMARY

Current problems—what we should stop doing:

- Looking at compliance as the best option for controlling risks.
- Letting our traditional management paradigms blind us to the true business risks.
- Making worker safety and health a low priority in our businesses.
- Using "bandaid" solutions to problems.
- Thinking "next quarter" in our strategic planning.

Fixing the system—what we should be doing:

- Considering current information and knowledge as measures of risk.
- Using *poke yoke* and patience to eliminate problems.
- Making worker safety and health a cultural value.
- Using long-term dynamic planning to plot successful risk-control strategies.
- Changing the way we think to meet tomorrow's business needs.

7

THE SAFETY AND HEALTH PROFESSION

We take a lot of things for granted. The freedom we enjoy. The economic system we live in. Having abundant food and materials to buy. Driving here and there. School for our kids. A solid form of government. Not having a political coup every week. Relatively stable currency. Not fearing the country one hundred miles off our coast. Rampant disease in check. Information at our fingertips. Being able to choose our profession or occupation. Staying up as late as we wish. Having entertainment options. Banks that don't go out of business and keep our money. Air that isn't polluted to the point that we can't breathe it. Being able to influence having a waste dump or freeway next to our house. Owning property. Having living quarters that are larger than 10 by 10 or not living with 10 people in the same room. Warmth or cooling by turning the thermostat up or down. Cold beverages when we are hot or hot beverages when we are cold. Not having to chop wood if we don't choose to. Having generations go before us that made all these things possible. Indeed we are spoiled. We take too many things for granted.

We take our professions for granted. After all, we have jobs. What else is there? What would happen if you were trying to get a job as a document counter? You didn't write them. You just counted them. Knowing how many documents a company has, you reason, is a critical piece of information for any business. You could have too many and that could be a real drain on the business. On the other hand, you could have too few and that would place the business in an extremely vulnerable position with the document regulators or the competition, or in danger of losing control of your processes or quality. This could be a very serious problem.

If you were to try to market your expertise in document counting, it would definitely be an uphill climb. First of all, you'd have to get companies to listen to your pitch. Second, you'd have to convince them of the need to have their documents counted. And third, you'd have to convince them that this is an ongoing business need. In other words, you would have to justify your profession. This says nothing, of course, about being offered a job in your profession with the company or how long you would keep it.

But what if the document-counting profession was already established. Let's say that there were some laws that regulated the number of documents that businesses had to have. Further, there were document number inspectors that would periodically, without warning, come into businesses and count their documents and if they weren't in compliance with the regulations, the document counters would cite them and give them monetary penalties. Let's say that there were areas of study where you could learn the fine art of document counting and even colleges that offered degree programs in it. Let's say that you could become a Certified Document Counter (CDC) if you passed some tests, and that made businesses know that you were proficient in what you did. Maybe there would be some professional journals on document counting where you could get current information. The profession could have grown to a point that there were even document counting consultants. Perhaps, the state in which you practiced was even considering licensing document counters to keep the riffraff out. Let's say that all document counters got together and began a society called the American Document Counters Society or ADCS. Wow! You would have created and legitimized an entire profession.

We take a lot of things for granted. Just like an actor entering from stage left, he or she didn't have to build the stage or write the play or create the demand that filled the theater, or anything. The only thing he or she had to do was to learn the profession, get the part, learn the lines and enter from stage left. That was followed by the easiest part—doing his or her job to the best of his or her ability. Like this actor, we learn our profession by formal schooling or the school of hard knocks. We get hired or moved into our position. We learn the specifics concerning the place or facility we serve and we enter from stage left. Too often, we just start our appreciation for our profession with our endeavor to do our best at our job and forget, or don't look at all, at what went before. Truly we take too many things for granted.

Finishing up my undergraduate education a number of years ago, I had no idea what I wanted to do for my working career. I knew that I liked biology, chemistry, and physics. Knowing little more than that, I couldn't even choose between them for majors. After all, I, like most, wanted to work less and make a lot of money. None of these subjects seemed to offer a life like that. If I chose chemistry, I could hide in a laboratory, in my little world. If I chose biology, I could play with the animals or study nature. I have to admit it, I'm not much of an outdoors person. In physics, I could play inside my own mind and see things that others wouldn't even be able to comprehend. Quite a choice. But, admittedly, I loved them all.

Having a wife and getting on in years, "we" chose for me to graduate as quickly as possible. Biology was the quickest. Chemistry and physics would become my minors. Odd combination. In my last quarter of school, I started to look for a career path. I began my search for a "real job," any job. I sent out a lot of letters and got one serious response. At a low wage, I began as an analytical chemist. No one seemed to be interested in my biology or physics. Nine months later someone asked if I would like to work for him half-time in the analytical laboratory and half-time as liaison with the work environment. What would I do? Oh, some seemingly silly things like take samples of air and reaction gases and analyze them to see what people would be breathing. That sounded pretty neat.

If anyone had advertised for an industrial hygienist back then, I wouldn't have applied because, why would I have wanted a career cleaning workers' teeth? This was at the beginning of industrial hygiene as a practice in American industry. The problem was that no one knew the profession existed. It was a very well-kept secret. So, I started as an industrial hygienist—more accurately an IH-in-training, before that title existed. The point is, the profession already existed. I just happened to fall into it. I didn't have to define any profession, it existed and was beginning to take a foothold in American business.

BEGINNING WITH HISTORY

The safety and health profession wasn't recorded as one of the Lord's creations in the Book of Genesis. Not even close. Industrial hygienists and occupational physicians trace the roots of their occupation back to Hippocrates and Ramazzini. Hippocrates was the same man who came up with the Hippocratic Oath for medical doctors. Sure, Hippocrates described some effects of occupational disease, but I don't think you can trace an occupation back to some published notes. Actually, the safety and health professions began, formally, much later than that. Say, a thousand years or so.

The Safety Practitioner

Let's start with the industrial revolution. Other than mining, the industrial revolution pretty well created the safety and health problems that have become prevalent in today's mechanized world. With the industrial revolution came a lot of new hazards to workers, and a lot of unhealthy exposures. These hazards produced what logic would have told us if we'd asked, a lot of injuries and fatalities. The Pittsburgh Study documented that. Then came the workers' attempt to gain compensation through the court system, and then workers' compensation. A lot of things happened back to back. In other words, the newly created hazards of the industrial revolution created a lot of demand for controlling costs associated with worker injuries and illnesses.

With the expanding demand, the professions of safety and occupational health also began to blossom. But because each discipline, safety and health, grew up separately, they had very different professional concerns. The safety profession began on factory floors and in the mines. It was literally a grass-roots effort. The beliefs that anyone could do the safety job and that the best spokesperson was a victim of a significant injury, bears testimony to this origin.

The fact that these beliefs still exist today shows how little the safety profession has advanced. Some of the major employers developed industrial hygiene functions in-house, but they were few. For the most part, perhaps 98 percent, the professions of industrial hygiene and occupational medicine began in research and in universities.

The different origins of these professions placed very different demands on each. There was the safety practitioner. This term "practitioner" is not used lightly nor is it used with the normal disdain with which the safety profession uses it. It is, however, the corner stone of the profession. It started as a grassroots, factory floor job—he was the "safety guy." He wasn't the safety professional. That didn't exist. Have you ever heard of an industrial hygienist practitioner? Of course you haven't. It's because of the very different origins of the professions.

In any event, because the safety profession began on the factory floor with nonprofessionals, the safety practitioner needed professional development. Having no formal education in safety, as opposed to health care, professional development was a constant challenge and always ran headlong into budget restrictions from the economic-based system. After all, safety is an easy thing to do, management would reason. Anyone can do safety; why should we invest in education for a job that anyone can do?

Consequently, the safety practitioner's job has always tended to be a transient occupation. Few found a love and stayed with it. It was not respected by management. Few at first, more today, chose safety as their lifelong profession. But, in any case, because of the origin of safety in business, safety practitioners have always been more closely aligned with the worker than management. This is an important point. It was also a result of sharing a common philosophy; the human, non-economic side of worker safety bound them together.

Occupational Health Specialists

Occupational medicine and industrial hygiene began with very different beliefs. Both were considered highly technical and required specific education and experience. Because of that, people who began in the fields tended to stay in those fields. Also, because they were much more hallowed professions and did not come from the grassroots origins, they have always philosophically aligned themselves with staff or other professionals, not with workers. Coming from research and medical roots, they also tended to be analytically oriented, focusing on data collection rather than on the human, emotional side of worker health. Different origins. Very different concerns and approaches. And different dealings in business.

Among those differences is the way each has dealt with and is treated by the economics of business. The safety profession grew up inside the economics of business as an economically-justified aspect. They were there to control the costs of injuries and workers' compensation. But because they were grass-roots in the organization, management traditionally cast them in this same economically-justifiable world. The safety profession has dealt with this cold economic world for their recommended actions and programs, and for their very jobs and staffing levels.

Occupational health specialists came into industry, for the greater part, long after the industrial revolution and the growing pangs of safety in America. They really didn't get a foothold in industry until the OSH Act created the need. Having documented need, they have always tended to deal with business economics very differently than their safety counterparts. For example, they think nothing about recommending a $100,000 local ventilation system. A safety person would be "drawn and quartered" for making such a recommendation. But, the occupational health specialist comes from the more respected scientific roots. No, it doesn't mean that their recommendations are any more embraced by management. That is still wrapped in the economics of business.

Management expects occupational health specialists to make such unrealistic recommendations. After all, that's what scientists do.

A COMMON VISION

The commonality, of course, is the vision. Both the safety and health professions started with one vision in mind—to improve the plight of the worker. That's an admirable goal, although many would say it was idealistic. Being idealistic myself, I can deal with having an idealistic vision. After all, isn't that what a vision is all about—having a lofty glimpse of what could be and striving to achieve it?

Setting aside for a moment that both sides came from different roots, the vision of where we want to go is very similar. Dealing with toes, fingers, arms, or legs or focusing on lungs, kidneys, or various systems such as nervous or pulmonary, each focuses on the whole worker. Healthy workers, now and 20 years from now. This is a noble, singular, ambition.

To a larger extent, both started and continued to move in an altruistic direction. Altruism is as near as it can be to being without boundaries or limitless. It sees no impediments, although there are many. It sees no limits, although technology places high walls around them. It sees no counter efforts, although they are continual. It sees only opportunity and discovery. It is indeed a wonderful, yet unrealistic world. No wonder each profession has had such difficulty dealing with issues such as licensing and professional fraud.

DIFFERING PATHS

The Science of Safety

Coming from very different roots, each profession has followed very different paths. Safety began on the factory floor. It was a humble birth. Not really knowing the "science" of safety, answers to questions came from very practical, try-and-fail methodology. That's where machine guarding came from. That's where fall protection came from. That's where electrical safety and lockout came from. They all came from very practical, shop-floor techniques to improve worker safety. It didn't require research. Try it and if it works, it must be right. Try it enough times and you develop new ways. More trials and you develop a safety process. It worked with exchange of information; sharing what worked and what didn't. It was basic, but the process worked well.

Over the years, however, the experimental side of safety has evolved. It has moved from the try-and-fail origins of the factory floor to much more sophisticated technology. Electronic and computerized control mechanisms are excellent examples. Look at the team-based aspects of system safety or process safety management as other examples of a much higher, more sophisticated search for safety answers. Sure, try-and-fail will always have a place in safety. As the issues become more difficult, however, the answer-seeking mechanism must also become more sophisticated. Try-and-fail techniques would prove too costly for robotic hazards. Computerized imagery and solution-searching techniques have proven invaluable. Looking at the problem-solving methods used by safety professionals, they are definitely moving toward more theoretical methodology. Not totally though, as there will always be opportunity for significant advancement at the factory floor level as workplace technology advances. Why? Because things get missed. Where else is there to finish the safe design work after installation? Nowhere but the factory floor.

The Value of Safety

The safety profession has also been trying to elevate itself away from its "common" origins. This has always been seen as a survival issue. With the "Safety Is Easy" paradigm so prevalent in American business, the only obvious way to counter it was a holistic, profession-wide inflation of the profession's worth. It is a change in perception-based movement. The best solution to this problem, of course, would be to make the changes at the place of the perception's origin, in management. Realistically, however, management's connection has never been a strong suit in the safety field, mostly because of its roots. So, a back door change of the perception of the profession seemed to be the most logical approach to this problem. In any event, there has been a continual effort to elevate the profession, to increase its level of respect. One of the approaches was to move away from the factory floor correlation to a more technical, theoretical connection.

This has been an effort of the safety profession. The safety practitioner (I use the word practitioner on purpose for emphasis) remains more connected with the worker. Still chasing the altruistic vision of the human aspects of safety, the practitioner serves as a champion in this movement. And on the factory floor with workers, continues to be the practitioner's most comfortable turf. So, the gap between management and the safety practitioner continues.

The Application of Safety

The occupational health profession, however, has traveled a very different path. Beginning in the research and university setting, they have been trying to move toward application. Research is one thing; application is very much another. This transition into industry, however, has been very slow. It is common for an industrial site to have a safety person, and depending on facility size, even a safety department. Even today, industrial hygienists and occupational medicine specialists are rare at most individual industrial sites. They tend to be more corporate staff- or business unit sector-oriented. These placements within industry make the occupational health aspects a specialty that is shared by, not dedicated to, facilities. This has had some negative impacts. The occupational health specialist becomes a hired gun, much like a consultant to facility management. Not one of us, but a spy from outside or an expert who creates unnecessary paperwork. Secondly, this outside staff orientation has not moved the profession toward any closeness with the worker. Occupational health specialists have become office beings, not factory floor, worker turf beings. They never have struggled to become comfortable in the workers' environment. They prefer the offices. Even their appearance supports this reality. Go to any meeting where safety and occupational health specialists are found. The occupational health specialists are always the ones in ties and jackets. The safety-types are in slacks and unbuttoned collared shirts.

Safety Professionals

Occupational health specialists have always thought of themselves more as scientists searching for the complete data set. They like numbers. This is precisely where the disconnection comes. The roots of occupational health don't allow an easy transformation to the industrial, economic world. I wish I had a dollar for every time I heard, "They don't understand what it takes to run a business" from a member of upper management after they had been "visited" by an occupational health specialist. Consequently, just as they don't feel comfortable as liaisons with workers, occupational health specialists generally don't do well with management either. They share that same problem with the safety practitioner. However, the safety practitioner deals very well, almost too well in the worker's world. Occupational health specialists, due to their roots and the reasons they chose the occupation, do not.

Coming into the profession of industrial hygiene from an unknowing position, I was an enigma. I never really fit in with the scientist-type occupational health specialist because I was learning in an industrial setting and didn't have the same roots. In other words, I was shunned by the scientific types. I was also a member of industry. That was my "second strike." Most occupational health professionals came from academia or government. I was always one of "them." At association dinners and parties, I sat by myself. I also didn't fit into the safety practitioners' realm either. I didn't spring up from the factory floor. I came from the chemical laboratory. I had a technical education. I didn't know a machine guard from a security guard or a clevis from a closet. I certainly didn't fit into their world. But, coming from worker-roots, I very quickly found allies and teachers in the safety profession that I couldn't seem to find in the occupation health associations. At least to the safety people, it was acceptable to admit that I didn't know and wasn't afraid of asking stupid questions. Grassroots—we shared those.

Over the years, I find that I am in an increasingly growing segment of the profession—beginning just before the boom times and not coming from the shop-floor nor the scientific community. We enigmas are becoming a large part of both professions, practicing in both arenas with competence and ease. The pure scientist and the pure factory-floor practitioner are disappearing.

COMMON PROBLEMS

Non-Alignment with Management

The non-alignment with management is probably the most common and destructive problem facing those who practice in the fields of safety or occupational health. They simply don't connect with management's way of thinking. Each of us has heard the battle cry "You've got to speak management's language" at least a thousand times, from those who realize this disconnection. Unfortunately, this "language" has been interpreted to mean "economic", which is wrong and has set the professions back, not forward. The disconnection, you see, isn't the language of economics, it's a way of thinking. Management deals with paradigms from management-past but they deal in options, risks, and costs. It is a near valueless world. If safety and health practitioners deal only in the realm of costs, they miss the most important area, values. They don't share common values with management. Not sharing common values, they can't share a common language. It's like talking to your pre-teenage son or daughter. There is no common value, therefore, there is no common language.

Reductionist Thinking

Secondly, the safety and health areas share a reductionistic way of thinking. It is their issue, their way of thinking. This single-issue, reductionistic way of thinking further retards any communication outside their arena. Why? Because others don't share it. Management, as an example, thinks more holistically. It depends on what's hot at the time, but they deal with many issues all the time. Safety and health specialists tend to get hung up in their single issue thinking that it should have equal or heightened value with others. It doesn't. This single issue generally doesn't consider financial vision either. Now many will argue differently, saying that inattention to certain exposures or hazards will break the bank down the road. Management, however, sees many things that can and have a higher probability of breaking the bank. Most, due to their holistic way of thinking, are indeed more critical to the continuation of the business.

Isolation

The third area where safety and health specialists have common problems is in the area of feeling that they are isolated. After taking a look at the other two common problems, it shouldn't be a surprise why. It's like walking down the street and meeting a person carrying a sign that says, "The world will end in 10 days." You are going to give him plenty of sidewalk. We would naturally consider him to be "from some other planet" or "on something." He would be isolated. Holding that image in your mind, picture a safety or health specialist who brings a single issue, "you've got to fix this now," or "we must spend the $100,000 dollars to remedy this problem now." Isolated? You bet they are going to be isolated.

This causes a more insidious and isolating result. Both specialties have tended to "circle the wagons" in an attempt to protect turf. Consequently, they tend to hide in their offices—the safety guy in the Safety Office and the occupational health specialist in the clinic or the Division Offices. This isolation further separates them from line management and the day-to-day realities and activities. Further losing touch is not a positive thing. But it is a common problem that is tying the professions closer together.

COMMON FRUSTRATIONS

I don't know of any occupation that has more frustrations. Grouped, they would include always working in a "fire drill" world, not being taken seriously by management, constantly having their programs and rules ignored by management and workers, never getting invited to important meetings or told about important changes in the planning phases, being paid less than their engineering counterparts, constantly not having enough staff or time to get their compliance activities done, lacking resources to make things happen, having responsibility for safety or health and having no authority over the day-to-day activities in the line structure, being called on the carpet for injuries or OSHA citations when management or labor didn't do what was recommended or set up—need I go on? The work occupations of safety practitioners and health specialists have always been highly frustrating.

It's like beating your head against a brick wall. You have two choices, keep doing it or quit. If you keep doing it, it is difficult to keep the sensitivity, purpose and dedication going. You tend to burn out, become desensitized and lose touch. So you usually revert to the second choice—you quit. There are, of course, two ways of quitting. You can quit and leave, perhaps to find an organization that is not aligned with this traditional American way or you can find another occupation. Both happen regularly. But second and more destructive, you can quit and stay. I've seen this happen too often.

Wouldn't it be nice to be given a dollar each time you see a sign marking frustration on the wall of a safety or health person's office? "I must be a mushroom . . . the way they keep me in the dark and feed me B.S." "Don't look here . . . I'm always the last to know." "If you can't dazzle them with brilliance . . . baffle them with B.S." "If you think OSHA is a small town in Wisconsin, you're in real trouble." "I don't know who I am . . . I only know that I'm not your priest and I'm not your mother." They can be called frustration signs. Sure, they're funny. But more importantly, they verbalize what we feel.

BLENDING THE PATHS

Industrial Hygiene

These professions are not pure sciences. They share, however, a common path that will continue to expand and blend. Safety practitioners are learning more about industrial hygiene. Some have even become certified in industrial hygiene specialties. As things have become more complex, safety practitioners have recognized that they need to acquire more scientific knowledge and skills. This, of course, ranges from a knowledge level of just knowing what you are talking about to proficiency in the specialty. But this movement has been gaining momentum and will continue to blend professional paths. This increase in scientific knowledge and skills seems to be removing some of the traditional ties to the factory floor and the worker. Just like the old saying about not being able to return to the farm after seeing the city, safety practitioners who learn the scientific parts of occupational health seem to move into limbo. No longer on the farm nor in the city.

Safety Practitioners

Health professionals are also becoming partial safety practitioners. But it's happening much slower and with greater resistance. Whether it is an occupational health nurse that learns more about lifting safety to stem the tide of back injuries or an industrial hygienist that learns about electrical safety to compliment his or her skill proficiencies, this movement at expanding knowledge and skills is a chosen path. It is a transition from a speciality-focus to a dilution of skill for one major reason—job security. The more you know, the logic goes, the more valuable you will be to the organization and less of a target in a RIF. This is causing an interesting result. In such workplaces, the safety practice is moving away from its traditional factory floor. The occupational health roots in the scientific arena, however, seem to be a significant confounder to the close-to-the-worker traditions of the practice of safety.

Along came a significant confounder to this professional path blending—environmental compliance activities. Actually, both professions have embraced the environmental area as their own. Realistically, however, it doesn't really belong in either. But environmental knowledge and skills are an important new area for both safety practitioners and occupational health specialists. Having this common confounder is bringing a notable blending of professional paths.

Workplace Environment Generalists

We've mentioned some of the reasons that the professional paths are blending already including job security and the increasing need to pick up environmental knowledge. There are other significant reasons. The most significant is, of course, the OSHA regulations. There isn't one book of regulations that are safety regulations and one that focuses on occupational health. The regulations hit on both sides simultaneously. The Hazard Communication Standard is probably one of the best examples of this blending. From a safety vantage point, standard NFPA warnings and precautions are parts. From the occupational health focus, MSDS and exposure guidelines become significant. And there are also environmental aspects listed in the MSDS.

Another reason is the always-present practice of the person in the safety and health role wearing many hats. As the issues and regulations become more and more complex, additional knowledge and skills are necessary. Additionally, the disappearing line between practice and purity is causing this blending. Take ergonomics as a good example. Is ergonomics a practice inside industrial hygiene because incompatibilities may result in occupational diseases like cumulative trauma disorders (CTD)? Or is ergonomics more a safety-related discipline because CTD is merely an injury in slow motion? Or, are they both wrong? Is ergonomics a practice within engineering? Does it matter? Most think not.

If you look at the aspect of moving up the ranks, blending paths is the name of the game. Moving from specialist to generalist and learning management skills is part of the necessary learning curve. If anyone plans on moving up in the ranks, blending paths is unavoidable.

Remember when it used to be simple? You were a specialist or a generalist. If you needed a specialty to work on a particular problem, you called in a specialist. Need a fire systems specialist? Call one in. Need a ventilation expert? Get one on the phone. It isn't that simple anymore. The workplace and worker safety and health issues are much too complex. Downsizing and budget constraints have and will continue to reduce resources. There simply aren't any other options. To practice safety and/or occupational health in today's or tomorrow's world you have to be diverse in your knowledge and skills. You begin to reach what I call the "peanut butter threshold." In other words, how thin can you spread knowledge and skills (and people resources for that matter) and still retain an effective level of knowledge and skill. Theoretically you can spread peanut butter into an infinitely thin layer based, of course, on particle size and emulsion efficiency. But, at one point, you lose the taste of the peanut butter. Therefore, you've spread it too far and too thin. Just like expanding our knowledge and skills, blending professional paths, at some point you reach your level of ineffectiveness. Kind of like a professional stretch to the "Peter Principle."

PROFESSION CONFUSION

Most professions seek to develop a collective voice. Usually this is accomplished via a professional society or association. Looking at the engineering, medical, and other professional segments of our society, they all tend to do the same things in order to gain a collective voice. The safety and occupational health professions are no different. Oddly enough, the safety profession, over many years, has always had just one major professional society, the American Society of Safety Engineers (ASSE). On the other hand, the occupational health area has many. Occupational nurses have one, occupational physicians have theirs, but industrial hygienists have several. They have the "everyone can join" American Industrial Hygiene Association (AIHA), the "only the certified elite can join" American Academy of Industrial Hygiene (AAIH), and the "only governmental- and academia-types can join" American Conference of Governmental Industrial Hygienists (ACGIH). The whole structure of the occupational health professions seems to be based on who can't join rather than who can. They tend to be elitist, exclusionary "voices," not a collective voice at all.

As with all professional organizations, they have many purposes. They are the collective voice of the profession. They look after growing the profession. They identify professional issues and develop a collective stance on them. They look at professional development and many look at professional quality control. This is where the certification aspects got their roots. These outgrowths, however, remain focused on what "hoops" professionals must jump through in order to become certified or keep their certifications. The professional organizations, however, have a much broader caretaker purpose for the profession. This is where the profession confusion comes in.

Mission Definition

Everything exists for a particular purpose. Using a business term, it is called the mission of that entity. A statement of mission simply tells why that organization, group, team, etc. exists. For example, Mothers Against Drunk Drivers exists to get drunk drivers off the road so they can't hurt anyone. The American Management Association's mission focuses on improving the practice of management in America. The ASSE exists to advance the practice of safety engineering and the AIHA focuses on improving and expanding the practice of industrial hygiene.

What's confusing about that? It all depends on how you define the mission. For example, let's say that my organization serves the profession of snail gathering and my professed mission is to advance the *practice* of snail gathering. We, therefore, can have two separate and very different definitions of that mission. We could define it to focus on the *practice*, that is, how the job of snail gathering can best be done. In other words, focus on gathering more snails, plumper snails, more efficient gathering efforts, etc. We could, however, define the mission very differently. We could focus on the "practice," meaning the *profession* of snail gathering. We could develop quality standards for those snail gatherers who could measure up to the organization's competency of practice. In other words, the definition of their mission could be focused on the "practice of the profession," not the "profession that people practice." These are very different organizational focuses. To accomplish either takes the organization down very different paths toward that mission.

Let's look at three real-life areas of confusion: profession mission-definition confusion, profession respect confusion, and profession specialty confusion.

Profession Mission-Definition Confusion

I tend to be an idealist. In order to fulfill my idealist tendencies, I try to simplify issues—take them to their most basic parts. To me, it adds a lot of clarity and removes a lot of the confusing shrubbery. This is the approach we should take to the practice of either safety or occupational health. The most basic question is "What are we trying to accomplish?" We could look at it from a very practical, hands-on perspective or from the very different holistic, big-world perspective. From this simple way of thinking, the big-world perspective must be secondary to the one-on-one practical perspective. The one-on-one is more basic. It, therefore is more important and causal. Okay, so what are we trying to accomplish? Aren't we trying to improve worker safety and health in America? This is the practice of our specialties. If this is true and all of us practicing professionals (and practitioners) are trying to accomplish the same thing, shouldn't this be what our collective professional organizations are trying to accomplish too? Makes sense, doesn't it?

Our actions speak a lot louder than all our words do. It's true of us and it is true of our professional organizations. If you focus on the actions of the professional organizations, it's a mixed bag. It's schizophrenic. At one time we act as if the practice is important. More often, however, we act as if it is the profession. This is where the confusion starts. Let me give you some important examples. All professional societies provide training opportunities. What training they offer generally represents the current focus of the association or society. If we classify the courses as fitting into the practice arena or the professional one, we usually get a 10-90 split, respectively. If the practice is so important, why isn't it a mainstay of our professional development courses? Why do we insist that you have to have a degree in your specialty, an "associated degree," or significant experience before we will allow you to be a full member of our professional organization? Are we trying to emphasize and encourage those that are learning the "practice," or accentuate those that are elevated in the "profession?" After all, by having layers or different membership levels, aren't we developing a caste system? What does this possibly have to do with furthering the "practice" of worker safety and health?

Profession Respect Confusion

We've all read a lot of position statements of professional societies and associations. They usually state that one of the highest priorities is to elevate respect for the profession. Why do we think this is important? Well, there are two possible reasons. The first is that elevating the profession will make "others" pay more attention to its professional groups. It's a "look at me, I'm important" issue. It says that if you are talking about a worker safety and health issue, you should consider the association's input important. Targeted at governmental actions, it seems to want a voice in the regulatory process. Guidelines are what will pattern tomorrow's world, not regulations. As a reason for seeking greater respect, this would be short-sighted and ineffective.

Personally, I think it's the other reason—a reason that is more power-oriented for each professional society or association. If we have more respect as a profession, then you, the practicing professional will have more respect in your workplace. This is, of course, total hogwash. If the National Accountant's Association (I made it up) has high respect in our society, does it mean that accountants will have high respect in business? No! Individual respect in business is earned on a place-by-place basis by what "value" the individual adds to the business. Even a recently developed "Code of Professional Conduct" adopted by one of the professional organizations charges members with "Striving to increase . . . the prestige of the safety profession." National respect buys you *nothing*, but our societies and associations don't seem to get the point. The point is this: Don't confuse the profession with the professional organization. The profession should focus on the practice of the specialty, not on the professional organization the members of the profession created. We need to focus on the practice, which is, of course, what we are trying to accomplish.

Profession Specialization Confusion

"It's a battle out there!" Everyone seems to be vying for what the others don't have. Its roots are in our society's competitive spirit, but when it comes to getting what others don't have or having it first, "it's a battle out there!" This is another area of confusion. If we are all trying to accomplish the same thing, improve the practice of worker safety and health in America, what does it matter who owns it? Take ergonomics as a good example. The occupational health associations rushed to take "control" of the specialty, thinking that it fit better in the area of disease. After all, where on the OSHA Log do you report it? The safety society also rushed to embrace it thinking that it was just an injury in slow motion. What difference does it make?

Environment is another example. Really, no national society or association exists that captures all environmental professionals or specialties. So, here come the competitive societies and associations. One makes it a specialty that you can get certified in. Another organizes a special society sector that you can belong to. There are too many associations and societies if all they can do is try to find a competitive edge. This just doesn't help.

We need to look at common ground that brings the separate professional organizations closer, not at specialities that each can find that distinguish each and takes them farther apart.

This whole phenomenon should be called "Mission Confusion." It's when you aggressively try to accomplish what you think you are trying to get done, but you're really headed in the wrong direction. In business, this happens all the time. Let's draw a direct comparison between the professions' "Mission Confusion" and what happens in business. It can be illustrated best using a training example. A training game used by Development Dimensions International (DDI) of Pittsburgh, PA, emphasizes this confusion very well. They play the game in training courses using four department teams that represent four areas of an organization. Each team is given two cards. One has an "X" printed on it. The other has a "Y." The object of the game is to "make as much money as possible and lose as little as possible." Each team is to play one card per round of the game. The scoring of that round is as follows:

Team Plays	Team playing Y score	Team playing X score
4 Ys	- $25	--------
3 Ys / 1 X	+ $25	- $25
2 Ys / 2 Xs	+ $50	- $50
1 Y / 3 Xs	+ $75	- $25
4 Xs	-------	+ $25

It doesn't take a rocket scientist to see the opportunity for gaining monetarily by being the only team to "throw a Y." Strategies are devised to set up the other teams, lull them, and then take advantage of them. This becomes the strategy in at least 95 percent of the training sessions. Through a practice round and through four game rounds, each team strives to accomplish what they think is the mission, make more money than the others, and lose less. When you stop, step back and think, that wasn't the mission at all. It was a misinterpretation of the mission. It was "Mission Confusion." The true mission was to act in the best interest of the organization, not each department team. Therefore, the mission is misguided, and regardless of which team thinks they won, the organization loses. This is "Mission Confusion" personified.

Applying this example to our professional organizations, what is our collective mission? Isn't it to advance the practice of worker safety and health? We suffer "Mission Confusion" and misinterpret the mission to mean "advancement of our professional organizations." By playing the game this way we, like those playing the DDI game, begin to delight in our perceived victories and not realize that we, as a collective practice, have lost.

CIRCLING THE WAGONS

The down side to "Mission Confusion" is that we take on destructive tendencies. It is a lot like animals in extreme stress situations eating their young. And not since the boom in occupational safety and health came due to the passage of the Occupational Safety and Health Act has this professional stress level been greater than it is today. All indications point to it continuing and becoming worse. These stress reactions are important to recognize because they are symptoms of our historical approach to change and, if left unaltered, they could very well have disastrous impacts on the professions.

What are these stress reactions? There are three significant stress reactions: use of short-sighted professional goals to rationalize importance; emphasizing differences, not similarities; and professional turf protection and isolation practices.

Use of Short-Sighted Professional Goals to Rationalize Importance

Since the first historical accounts of warfare and territorial power, numbers have always equaled strength. From the Israelites versus the Philistines, in which David and Goliath took on significance, to World War II accounts, the number of "warriors" one side could mobilize always was a measure of strength. One important flaw in this thought process, though, was that the biggest group didn't always win, nor did the best trained or most specialized. Look at the Revolutionary War. Our professional organizations, however, have always believed this paradigm. The constant membership drives and pushes to have annual conference numbers be the biggest ever are testimonies to this way of thinking. We don't deny that more membership is somewhat a measure of a profession's growth. But, in fact, it's always meant more than that. It has been a measure of clout, power, potential influence. Unfortunately, it isn't a good measure of this at all. One only has to look at the increasingly transient nature of the growing membership to recognize this.

Continuing to emphasize numbers can delude the profession. This is especially true when these numbers are used to rationalize current and potential negative impacts on individual professional members. How can you, the membership, be concerned about changes in your workplaces that might "displace" you and your services? Our ever increasing membership and conference attendance indicates that our profession is healthy and not at risk. Of course the current "streamlining" trend is a concern to those who practice the professions. It should be. Job security literally equates to food on their table and shelter. These concerns are too basic for any professional organization to rationalize away and not focus active resources on dealing with the membership's concern.

Incidentally, our professions aren't the only ones to display this type of protective behavior. Look at American business and how they emphasize short-term numbers, the next quarters financial results, rather than creating a vision for the long-term future.

Emphasizing Differences, Not Similarities

This is another stress reaction, a circling of the wagons. Psychologists tell us that it is totally normal to do this in a stressful situation in which we conclude that we, as individuals, are at risk. They tell us that by emphasizing our special differences and our abilities, this elevates us above the target group and we blend more into the group that is not at risk. I'm reminded of a Gary Larson cartoon in which two bears appear standing, being seen through what is obviously a rifle's scope. One bear appears unaware of the danger while the other, smiling at the hunter, arm around the unaware bear, is pointing at the other bear as if to say, "Take him." This is a good example of this normal stress reaction.

How does this stress reaction appear in today's professions? Actually, there are hundreds of examples, but let's look at two of my favorites. Some of us remember when there was only one professional certification. You were either certified as competent to practice your profession (safety, industrial hygiene, occupational nursing, etc.), or not. Now you can be certified in a number of specialties including management, environmental, and a number of other areas. Ever stop to wonder why we have so many specialties that you can get certified in? Often we think that it is a way to sort the "wheat from the chaff." More importantly, it is a tool that professions use to claim turf and emphasize differences through specialties. If it weren't, it would be used to limit professional practice, like medical doctors. Look at other practice certifications that are used to focus on competency to practice. Look at the CPA, Certified Public Accountant. Are there CPAs that are certified in tax accounting or just in personal finance? Are there CPAs that due to their certification can only practice in government or business? No, because they don't have specialties. The CPA stands for competence of practice only.

Maybe all these professional specialty certifications the safety and health professions offer are merely there so that we can tag more letters on the end of our names as printed on our business cards. Just look at all the alphabet soup that we use to follow our names; it makes one wonder if the letters aren't more important than who we individually are and what gifts we have. After all, it does stroke our egos, reinforce the caste system, and emphasize our differences.

What if we had only one common certification? You were either certified as competent or not. This approach just wouldn't work with our current paradigms concerning the specialties of practice that separate us from all the others. It runs counter to our "we are special" (meaning separate and proud of it) way of thinking. Benjamin Franklin, perhaps, said it best at a rather significant and stressful time in our nation's evolution. He said, "We must all hang together, or most assuredly, we will all hang separately." What is so eloquent about this statement is that it speaks to us today. If we do the natural thing in a stressful environment by separating ourselves from the other practices, we set our specialties up to be "hanged." Look at the practice of industrial engineering if you want an excellent case example from history where a profession almost became extinct because of too much specialization and isolation.

The second example of stress reaction is the caste system that we have created in our professional organizations. For instance, look at the multiple associations that the industrial hygiene profession has. Each group sets itself apart from the others by who is allowed to join and who is not. It is an excellent example of emphasizing differences and not similarities. Another example is the levels of membership that major professional organizations like the AIHA and the ASSE use. You have to "measure up" by education or by practice if you hope to be granted full membership. This practice builds an internal caste system of those who "meet the mark" and those underclass persons or non-professionals who can't be part of the real club.

Professional Turf Protection and Isolation Practices

In stressful environments, we humans have built forts, castles, walls, moats, and in business, we build offices. Actually, they are all forms of turf protection. There are more subtle forms of turf protection that we also actively practice. These may include approval of drawings or procedures, being the only ones to do accident investigations or area compliance inspections, being the approving office for safety equipment, being the only ones that can take noise measurements at hood velocity tests. Our subtle forms of turf protection are only limited by our creativity. But they are definitely forms of turf protection, just like walls, forts and moats.

Turf protection is normal. It combats our insecurities to believe that our employer or others cannot do without us. In fact, they can do without us, and with the "right-sizing" trends in business, they are discovering that fact quickly. Turf protection is fine, but it is usually temporary at best and only counters our feelings of insecurity. It is a protective strategy, not a contributing and growing strategy at all.

Unfortunately, professional organizations mirror this turf protection practice. And, as this practice isolates us in our professional organizations, it also isolates us individually in our work environment.

ADVANCEMENT THROUGH "PRACTICE"

Okay, so what do we do to advance the professions? What can we learn from our past and our current behaviors to help us advance our mission? There are countless opportunities, but these should really be derived and implemented by mass membership input and action. Let's discuss two simple ideas.

Clarify the Mission

We need to be reminded as practitioners and specialists, and as professional organizations, what our common mission is. We must be dedicated to advancing worker safety and health. Focusing on this mission, we need to honestly evaluate what we do individually and as professional organizations that does not compliment or move us toward this mission. These would include our turf protection and isolation behaviors and our search for professional "respect." This honest evaluation will require great candor and courage.

Focus on the Practice

If we are clear about our mission, it really doesn't matter which specialty does what, only the quality of that service. We get too hung up in the profession and our highly regarded specialties and forget that it's only through the practice that we can advance our mission. It isn't a power trip. It is about the practice of the profession.

More troubling is our habit of taking behaviors and professional organization direction and policies as gospel. We simply don't question them. We take it for granted, without evaluation, that they are sound and that, if continued, the future of the professions will be assured. This assumption is extremely dangerous, especially in a world that is changing as much and as quickly as ours is today. Knowing that changes will come faster and be more radical in the future, stop and think about the impact tomorrow if we don't make some changes now. Now is the time. If we choose to ignore what we see today and not make the necessary changes, we may not get that opportunity tomorrow.

SUMMARY

Current problems—what we should stop doing:

- Focusing on the profession rather than the practice.
- Trying to keep our turf and resist change.
- Resisting learning new skills and becoming more multifunctional.
- Reinforcing the professional caste system

Fixing the system—what we should be doing:

- Clarifying our mission and dedicating ourselves to advancing worker safety and health.
- Focusing on the practice, not the profession.
- Getting actively involved in change so that we can help form our tomorrow.
- Dismantling the professional caste system.
- Unifying our collective voice.

8

THE CHANGING FACE OF BUSINESS

RATE OF CHANGE

Have you ever wondered what it would be like to live in a time where things were slower, more assured? Before the industrial revolution, changes focused on seasons of the year and what you were wearing. Even what crops were grown didn't change that much. It wasn't until the early 1900s that we learned that crop rotation didn't sterilize the soil. The weekly newspaper (if you lived in a populated area) talked about the baby the Jones' had or about the church social that was a success. Times moved much slower then. Will Rogers, an early 1900s humorist, told about watching the corn grow as a form of entertainment. Life was simpler then. Change was more gradual. But I doubt, even at that slower rate of change, that it was easier for the people of that time to accept change with grace.

The rate of change has definitely changed. Today we deal with change almost daily. Our normal route to work is under construction or has a traffic snarl, so we take an alternate route. Our great grandparents took one route home or to church and never had to change it. We struggle to decide what to eat for dinner. Should it be chicken, pasta or should we call for pizza? Meals at the turn of the century were very unchanging and, depending on whether you lived on a farm or in the city, the diet was pretty well set by what you or your local area could produce or grow. Today we choose what to wear daily. Should it be the black suit and the blue tie or the pinstripe with the red? Back then, you had one suit, or two if you were affluent. If you lived on a farm, you wore the same bib overalls until wash day. Change was not a day-to-day thing then.

Magnitude of Change

Today, change comes in greater magnitude. Computers today are radically different, cheaper, and more powerful than just ten years ago. New technology is available and more affordable each year. Cars come with CD players and independent and four wheel steering today. Many cars on the road still have tape players and rack and pinion steering. Correspondence that took a week or two just a few years ago can now be received in a matter of seconds via Internet or fax. People in the early 1900's couldn't even fathom the magnitude of changes we take for granted in today's rapidly changing world. As a matter of fact, we expect it.

I remember a saying that was used when I was growing up, "Same old . . . same old." It meant that what you were expecting at work, at home, or elsewhere, was just as you experienced—unchanged. In today's world, no one can use this once-common saying. Perhaps we should say, "Same new . . . same new." In fact, we never seem to know what we are to expect. Change is not just a fact, it is a part of our daily lives.

Growing up in southeastern New Mexico, the wind seemed to never stop. The wind was always in motion. If the wind ever stopped, we'd exclaim, "What's that?" The mere fact that the wind was not in motion would signify a significant change worth noting. This is exactly like change is today. Change never stops, it is always in motion. If suddenly, somehow, it were to stop and things remain unchanged for any period of time, we would just as assuredly exclaim, "What's that?" The lack of change would be deafening.

Business' Response to Change

Business too is caught in this upward spiral of change. Business has evolved to a rapidly changing world from one that didn't change much at all. Following the industrial revolution, America formed not only its industrial base but also its business structure and practices. As the businesses got bigger and more profitable, the basic structure and practices changed little. Management was autocratic. Management and labor were divided by roles—management thought and labor worked.

Manufacturing was the mainstay of our industrial base. Man production manufacturing techniques that were born in the Ford factories grew and were refined, but changed little. The demand for basic metal and the methods by which it was produced changed at a snail's pace, although the size of the production facilities became larger. Businesses grew, but they changed very little. More importantly, business practices changed even less.

World War I further entrenched these practices in American business. Obviously, these practices were successful in supporting the war effort, just as the engine of business was successful in supporting the war needs of World War I, so it was in World War II. Successful again, it became even more entrenched in the American business culture. After all, America was the victor. We obviously had the best support mechanisms in business. Why then should they change?

Throughout the fifties and the sixties, our business experiences worldwide would support this belief. America's business methods and practices were the best. Remember the story of the tortoise and the hare? Let's look at today's version of that old story. At the start of the race, the hare streaked off leaving the tortoise far behind. The tortoise, however, recognized that she could not hope to keep up with the hare given her limited technology and his physiological advantage. So, being creative, she constructed a rocket sled upon which to ride and catch up. It worked magnificently: the tortoise caught and passed the hare at about mile five. What did the hare do to reverse his suddenly losing ways? Were the steps the hare took effective and did he regain the advantage? Who won the race? Well, that's what this chapter is all about.

THE AMERICAN WAY OF DOING BUSINESS

Axioms of American Business

Right Until Proven Wrong. Americans have always been firm in their resolve. This is a nice way to say that we're stubborn and bullheaded. We tend to think that we are right and that others aren't, until, of course, we are proven wrong. Even then, we show amazing resolve in discounting the proof, holding on to the American Way. It isn't one of our most admirable qualities. This has been the case in our foreign policy and our domestic policy. It has also been this way in our business methods. From a business perspective, it has been the result of "if it isn't broken, don't fix it" way of thinking.

We Want It Now. Americans have another not so endearing quality that is manifested in business. We aren't patient. This is why driving on our freeways is like living in a war zone. Our philosophy is simple, "Do it now, fix it now, and I want it now." It's the way we deal with problems at work, with our kids, and our possessions. It's the reasoning behind our love with and abuse of credit cards. If we have to wait, we simply aren't interested. Economic experts continually say that Americans don't save enough and when they start, it's often too late.

What's Best Is What Is "In". We also tend to be fad chasers. It matters little if it has to do with clothing, cars, who we know or associate with, what magazines or books we read, what or where we eat, who we listen to, or what we believe, we chase what's "in" all the time. If it's "in" and if we can afford it (or stretch our credit to "afford" it), we simply have to have it. Business is no different. If it is marble entryways, mirrored glass exteriors, or corporate jets, American business chases fads all the time. This extends from business logos to management fads. If it's "in," chances are that American business will try to have it.

Top-Down Structure. American business is traditionally top-down in its philosophy. Power comes from the top, as does vision (if an organization vision exists), and resources. There has always been, with few exceptions, a wall that existed between management and workers. Traditionally, we have called the people on either side of the wall white collar and blue collar. Are you a worker-bee or are you climbing the corporate ladder? Are you a wage earner or are you salaried? We've already talked about this traditional barrier and role separation between management and labor.

Effect on American Business

How do these four non-appealing qualities of Americans and American business affect changes in business? First, if it isn't born in America, we tend to discount it as having any value at all ("not invented here.") Therefore, business and management concepts that are successful in other nations become more curiosities than serious considerations. If something is obviously good and works, we tend to put our own twist to it and Americanize it. Then, of course, it becomes a whole lot more acceptable to us. Second, because our way is the right way, we tend to be slow to catch on to obviously successful business strategies that aren't American. This was why we were slow in joining the quality revolution. Third, if new ideas require taking time to implement them, we lose interest very quickly. We don't seem to be able to get past the next quarter's financial results, our do-it-now expectations. Fourth, like in our private lives, we tend to chase fads in business. This becomes even more complex when we try to turn effective management or business concepts into fads. But, of course, it allows us to Americanize them. Too often, our Americanization adds too much of our traditional business baggage, so the fad doesn't work. Being impatient, we then quickly move to the next fad. Call it the "plan of the week." And fifth, when ideas run counter to our top-down management philosophies, we have great difficulty warming to them. After all, our top-down approach has been successful for so many years and the barrier between management and labor is an institution of the American business establishment.

THE TRADITIONAL FACE OF AMERICAN BUSINESS

No one can argue the past success of American business. For a long time American business dominated the world economy. Compared to today, the world economy wasn't much, but when compared to what the Americans had, there was no contest. There's an odd thing about running second, third or last—you've got nothing to lose. There's something that is more odd about American business. When we were number one in the world, we didn't hear the footsteps behind us that were making up ground. But a lot of the reason for this can be discovered in an analysis of American business' traditional face while they were number one.

"American Business Is Number One"

Winning two wars certainly established American business as premier in the world. We were the only major power that was untouched by World War II's destruction. Business worked so well that it could support war efforts in two separate hemispheres of the earth. That, you must admit, is pretty impressive. After the war, what reason was there to change the traditional ways business had used to become successful? There was no reason so it became deeply entrenched in our American business culture. It was autocratic. It was economically focused. It believed that bigger was *always* better. Its structure was deeply invested in the organizational pyramid where power and direction flowed down the line to the workers. Very little percolated up the pyramid.

The Loyal Consumer

There was a traditional thought process in American business that was prevalent and unquestioned. Worded, it would sound something like, "If we build it, they will buy it." They, of course, were the consumers—key word, consumers, not customers. This thought process was a natural outgrowth of the traditions embodied in the organizational pyramid. We (management of the business) will tell, you (the employee or the consumer) will listen and do (work or buy). An autocratic dream based on supreme power. Needless to say, it wasn't exactly a customer-focused way of doing business. Consumers were seen only as those who would eventually buy the product. They were definitely not seen as discriminating consumers, nor did they have opinions that the businesses were necessarily interested in meeting. We produce it, you *will* buy it.

There was an obvious need at this time to be able to dictate to the consumer what their desires were, to create the consumers' desire. The expansion of mass communication and advertising made this possible. Donna Reed, the Cleavers, and all the other shows also firmly entrenched consumer expectation of what American life was supposed to be. This was a powerful tool for creating demand.

"Bigger Is Better"

"Bigger is better" was also a dominant business thought. Regardless of whether you refer to the size of businesses, cars, homes, or whatever, if it was bigger, it was obviously better than before. Here came the cars that are today known as "boats," with big bodies, big and bold lines, lots of lights, seats you could almost lie down in, leg room galore. IBM and Univac began marketing bigger and bigger mainframe computers. Airplanes began to not only get faster but, more importantly, bigger. It was obvious that bigger was better than faster when the SST (super sonic transport) idea was "shot down" and the Boeing 747 wasn't. And pushing this idea of bigger is better onto the consumer, the businesses got bigger, and bigger and bigger.

This firmly entrenched traditional philosophy and bigger and bigger business made "change" virtually impossible. Set full-steam ahead on a course that the captain was determined was right and successful, it became nearly impossible to turn the ship even a little. So any process or production change seemingly took forever to get done. This included organizational changes also. Change was not easy or rapid. Most of the time, it just wasn't done. Business was obviously successful, so why change it? A generation of business dinosaurs was created, slowing plodding through the swamp of consumers toward extinction.

How Do You Eat an Elephant?

"How do you eat an elephant?" "One bite at a time." Looking at the picture of American business, stuck in their traditional ways, this simple joke bears a poignant message. A leading economist put it this way in the early 1980s when it was obvious that America's businesses were no longer number one in the world: "We were entranced . . . we never heard the footsteps." That's one of the problems with being number one for a long time: you begin to delight in yourself and quit being attentive to what is happening all around you. That's what happened to American business. But being shocked awake from our pleasant nap, we had no trouble recognizing the obvious. We were getting our pants kicked!

Let's take a lesson from history. Question: which nation dominated the industry of fine watch making in 1978? The Swiss did. The watch-making industry represented more than 40 percent of the Swiss manufacturing sector. It employed over 100,000 people. A Swiss-made time piece was considered the finest that was available. Second question: which nation dominates the industry of fine watches today? No contest—Japan does. What happened to bring about this drastic change? Well, it was two things—a technology advancement and a paradigm. The technology advancement was the quartz movement. No mainspring, it doesn't require winding, has minimal moving parts, is dependable far beyond the mainspring-powered watches, the cost is less, and it is much more accurate. You are probably wearing this new technology on your wrist as you read this book. Needless to say, the quartz movement has been very successful.

What about the paradigm? The Swiss had been so successful in the old paradigm, the mainspring watch, that they couldn't see the new paradigm, the quartz movement, coming. They didn't even consider it. When the idea was presented by their own watch engineers, the Swiss discounted it. They were so successful in the old paradigm that they could not see the need to change. Their paradigm was so entrenched that they didn't even protect the idea! History tells us the results of the Swiss missing this opportunity to change. To the Swiss watch making industry, the results were disastrous.

How does this example apply to us today? Like the Swiss, America industry became too successful in their paradigm, the American way to run a business. We had our own people tell us differently, like Demming and Juran. We didn't listen until our paradigm was under siege in the expanding global economy. We didn't hear the footsteps of others catching up, running along side and passing with ease. But we are not talking about a certain industry here. We are talking about an entire industrial base—everything. Some businesses made changes early. Others took notice, but the largest segment still didn't get it. They saw the need to change but, like the Swiss watch-making industry, were too invested in the traditional way of thinking.

Change became a buzzword to us. I say buzzword because recognizing we needed to change wasn't enough. We had created a slowly moving, uni-directional dinosaur that was not easy to move, change its size or speed or change course. It was, in fact, a matter of our own creation. It was honed and refined just the way we felt it successful. Changing it now would be a monumental challenge. But change we would have to if we were to compete and survive in this expanding and improving world economy, and if we ever wanted to be number one again.

THE EXPANDING AND CONTRACTING BUSINESS WORLD

American business had many choices to use in this reinvention. More realistically, because of the dinosaur and traditional philosophies that were so entrenched in our culture, we only had a few. The rule was that the ideas would have to agree with American business paradigms and the way our business cultures worked. This is one of the reasons that management methods such as total quality have struggled in American business. They are simply too foreign to gain a foothold in the culture. They would require too much change up front in our philosophies and paradigms before they would stand much chance of producing organizational or business changes.

So, still economically focused and left with obvious problems on the balance sheet due to decreasing market shares, American business chose the only path that was traditional, one that would be accepted in management and would have an impact. Whether you call it downsizing, corporate streamlining, or re-engineering, it is the same traditional way American business has dealt with balance sheet discrepancies in the past—just whack away until the numbers balance. Actually, looking at American business over the past twenty years, it has been very cyclical. Whack away until the organization is too lean, a state of corporate anorexia. Then grow it slowly until the balance sheet suffers again. Downsize again, slowly expand, re-engineer until you again reach corporate anorexia, then slowly begin the growing cycle again. It's cyclical. Learning nothing from history, the cycle continues again and again.

Impact on Staff Organization

Bearing the brunt, percentage-wise, of these changes in American business is middle management and particularly the staff organization. It has been impacted to a great extent in relation to its power base. Those functions of the staff organization that have a strong power base, especially where it is supported by longstanding business-wide acceptance or stereotype, get impacted less than those that don't. One of the areas where this re-engineering effort has been particularly high in impact is in the staff functions that support the safety and health of workers. Historically, as safety and health functions have been downsized, they have been gradually rebuilt via the slow growth part of the traditional business cycle. This time, however, it's different. Before, the threat of regulation and OSHA, and the growing number of regulations caused a rebirth of staff functions. With new ways of dealing with these concerns, a de-emphasis of OSHA, and the blurring of lines between legal requirements and liability risks, the boom part of the cycle for safety and health is going to have significant difficulties getting started again. In the opinion of many, it won't happen.

In American business, safety and health has traditionally been seen as an economic tool, a form of cost avoidance. Placed in this light, it has always been a necessary evil to business. A business has always had to have the function, not because they wanted it or saw any real connection to the credit side of the ledger. In a downsizing effort in today's world, these economic connections and "have to have" paradigm form a significant weight about the safety and health function's neck.

The safety and health function has also been used as a tool in the negotiation process with organized labor. Let's face it. Organized labor doesn't have the clout they used to have. Look at Caterpillar as a textbook example. It doesn't take a rocket scientist to see that management has returned firmly to the "me management—you labor" autocratic style, driven by our current re-engineering efforts. And labor simply isn't in the power position to do much about it. This is another negative when it comes to the upswing of safety and health staffing following the current downsizing effort.

A FAD-CHASING HISTORY

Like most Americans, American business has always chased fads. This fact invites two interesting questions. What causes business to chase fads? And, why do they create fads rather than apply useful new concepts?

What Causes Business to Chase Fads?

The list of management fads that American business has chased over the past twenty years is very impressive. Those in the organizations call it the "direction of the week." They never stay. They get a lot of management verbiage, posters, slogans, cost money, and then disappear as suddenly as they appeared, only to be followed by the next fad. There have been self-directed work teams, quality circles, coaching, employee empowerment, benchmarking, total quality, verbatim compliance, employee empowerment, leadership, time-based management, cost-based management, product-based accounting, on-line purchasing, just-in-time manufacturing and invention, qualified vendors and suppliers, metrics, key competencies, pulse points, etc., etc., etc. Why does American business chase them? If they really aren't serious about instigating a change, why go through all the effort and expense?

The reason is a lot like the reasons we chase fads individually. When a new head or shaft for a golf driver comes out, why do we have to go out and spend $500 to own one? The reasons are both internal and external. How do you feel when you drive that new car or take that new driver out of your golf bag? Pretty good, don't you? This internal reason for chasing fads, although usually short-lived, is a significant reason why we always get caught in the fad's web. It just feels good. The external reason has a lot to do with how important it is for us to have the approval of others. We don't want them to think that we're not cool. It simply means a lot to us.

American business is no different. Each business exists in a society of businesses. To be among the first to embrace a fad, or at least to try, it is equated directly with being on the cutting-edge of the business world. Just as important, investors also read it that way. In case the fad works, we don't want to let our competitors be the only ones to benefit. Being caught up in a fad also gives us the right to use the buzzwords when we talk. Hey, that's important too.

Why Does Business Create Fads rather than Apply Useful Concepts?

Actually, if you look at a list of fads that didn't really stick to the Teflon surface of the American business culture, there are some really good business concepts. Some that have been very successful. Why have they only been considered fads within our business culture? It's happened this way for seven reasons. First, we have a greased mechanism for fad introduction in business. It's just too easy to insert it into the fad mill. Actually, it has become difficult for American management to recognize the difference between seriously considering a new concept and a fad. Second, some concepts run too counter to our traditional ways of doing business. For example, the paradigms of our business culture hold to a top-down, highly autocratic, management-thinks-and-workers-do philosophy. Some concepts are just the opposite and would require basic, fundamental changes in our culture before they could be anything but a fad. Third, we have a bias in this country. Call it the "Not Invented in America Bias." We simply think that if it isn't invented here, it can't have much value. So concepts from foreign business are naturally treated as fads until we either Americanize them or are convinced that they have value. Fourth, American business traditionally thinks in the short-range. We are continually focused on the next quarter's financial results. Concepts that are long-term run counter to our traditional short-term thinking. They are simply not quick enough to meet expectations. So they are inserted into the fad mechanism. They are never given a chance. Sixth, in America we have an innovation expectation. Innovation is a warm word, one that in America means radical change that immediately changes everything—success from failure. It's a "white light" concept. Concepts that are more grass roots in nature and involve small incremental improvements, like the Japanese concept of Kaizen, or continuous improvement have a hard time meeting our innovation expectation. They get sent down the fad line. And seventh, we have no patience for the long-haul. Recognizing that other businesses are far ahead of us, we don't want to go through the same learning curve they did. We want to jump to the head of the line instead of going through all the steps. It's like an instant formula: all it needs is to just add water—poof! The simple fact is that we are too far behind. Many of these new concepts we are trying were tried and made successful by others over twenty years ago.

There are excellent examples of these powerful business concepts that have been dealt with mostly as fads in American business. Concepts like just-in-time manufacturing that began at Toyota some twenty-plus years ago, total quality that most businesses still don't understand, and time-based management (that was also begun in Japan twenty years ago) are excellent examples. Some businesses have been able to use these to radically improve and compete in the global economy. Most, however, have used these only as fads and have clung to their traditional ways. It's the dinosaur march to the graveyard.

Why have concepts like re-engineering been accepted much more widely in America? Well, it's an American tradition. Whether called downsizing or some other term, it is a top-down, autocratic method of trimming rather than improving. It is a results-oriented, short-term concept that has immediate results. It is not process oriented or long lasting. Both of these concepts are foreign to American business.

THE EMERGING ORGANIZATION

Make no mistake about it—American business is changing. It has a choice—change or perish in the global economy. But, whatever mechanism is used to implement change, there are some predictable outcomes that are important to safety and health efforts. The traditional organizational pyramid is in danger of extinction. There will always be an organizational top structure and the organization will always rest on the worker's shoulders, but the shape of that structure will be vastly different. It will be different in two ways. The line structure will be flatter and there will be less support staff.

Flatter Structure

The cost of supporting the traditional multi-level management structure is simply too great in today's highly competitive world. This will only become more restrictive in the future. Management will have to do more with less and do it more wisely. This alone will force a lot of changes in traditional paradigms and roles. Workers will have to take up some of the slack and accept more responsibility (and accountability) if their respective companies are going to be successful and not fold. This flies in the face of union contracts and management concepts that are firmly entrenched.

But their change to a much flatter organizational structure is inevitable if survival, much less success, is going to happen.

Less Support Staff

This is both a result of questioning whether this highly competitive business structure can afford the traditional support staff and whether there exists a need for it in today's and tomorrow's workplace. These are very fundamental questions to safety and health programs that have traditionally been staff-supported. From a "can we afford the staff function" perspective, this is directly in line with the original economic, cost-avoidance role in which safety and health have always been placed. It is one more convincing argument against staffing based on newly recognized costs. After all, fears of higher workers' compensation costs and OSHA fines fall suddenly silent when compared to the need to reduce costs in a "survive or die" economic climate.

What will the results of this smaller support staff be on American business? Well, like the flatter organizational structure, more is going to have to be done with less, and more wisely. Obviously, line management and workers are going to have to take more responsibilities for the functions that are served by dwindling staff resources. From the vantage of worker safety and health, this is particularly good. Line management will have little choice but to accept that safety is a line management responsibility. For the safety and health professionals who have clung to their protective turf, it isn't a pretty picture. There is already a movement to out source a lot of the specialized safety and health functions that are difficult or impossible for the line structure to absorb. This is only going to expand in the future.

The writing on the wall says that if your staff function is dependent upon economic, cost avoidance justification, look out! There will be no room in tomorrow's business world for waste, including economically-based staff functions, when new, more poignant economic challenges are thrown into the equation. Staff functions that are or become value-adding will have much greater staying-power in tomorrow's business.

ARE THE CHANGES GOING TOO FAR?

We in America have another tradition. We never seem to stop in the middle. The pendulum swings from one side to the other, then back again. Like watching a tennis match, we never seem to find middle ground. So, it's a fair question to ask, "Are the changes going too far?" It's an important question to ask in that these changes in American business are only beginning. But, most assuredly, they will move through the center to the other side of the pendulum's reach. Swoosh!

To consider such a question, it is important that we take a position that is not too close to our perspective. Stepping back is difficult when issues like that tie so closely to our self interest are dissected. But it will be critical if we are to be fair in answering this crucial question. It is indeed too easy to get mired in a single-issue argument and see things only from our close-up perspective. A more distant, holistic perspective is required when seeking answers.

Stepping back, let's look at the possible answers from five different perspectives: the American economy, American business, the organization, the worker, and the practice of safety and health.

The American Economy

The success of our country, to a major extent, is dependent upon the success of American business, not foreign firms that purchase American resources, not foreign companies that build factories here and ship the profits back to their home country. Our success as a country depends on the success of *American* business. It's what our economic system is built on. It's what our tax structure is dependent upon. It's not just critical, it's much, much more important than that.

American business *must* change if it is to survive. The questions become: can we change quickly enough, and will we choose the right ways to change? These, of course, are impossible to determine this early in the game. But our only chance is if they do successfully change and change quickly enough to compete. We cannot depend on governmental actions to protect our businesses as we have so many times in the past. These are delaying tactics at best. In fact, governmental actions have no chance to counter the countless and rapid changes that are occurring. Only business can change quickly. But we need to develop that changing culture.

Bluntly, if our economy is to survive and support us and our children, change in business must happen!

Phases of American Business

The changes that are impacting American business today are mostly reactionary changes that are economically based. It's an attempt to position business so that it can compete. The changes are, therefore, in their infancy and they will happen in two phases; each phase will have successful companies and those that will perish. The first phase is economic posturing. This is the phase we find ourselves in today. Posturing is always more painful because it is the first and most cruel challenge to the traditional organization and paradigms. Are we going too far in this posturing? Not if our businesses never reach the point where they are postured to begin the second phase of change. So any change that moves American business toward this goal is essential.

The second phase is a lot more murky. It is the phase where we begin to reinvent American business into what it will be in tomorrow's economic reality. What will that look like? A lot of futurists think they know, but in reality, they are only guessing. The only thing for certain is that it will be radically different from our business organizations of today.

Will some parts of the American business community go too far in this transformation? Of course. Some will also not go far enough. Both extremes will place them at critical risk for survival. The majority, many of which we don't even consider major players today, will change into key players in tomorrow's world business economy.

So, is American business going too far? The only danger is not going far enough in this first phase of change. Being in a proper financial posture is the only place business can be in if it is to move into the second phase of change necessary to insure survival tomorrow. Can we go too far in this reinvention phase? Probably, but until we complete phase I, we don't even know what the playing field looks like, much less what shape the ball is.

Organizational Structure

Organizational structure is changing and will continue to change. But what we are seeing today is only the beginning of this first phase change. Realistically, it would be un-American if it were done any other way but in a top-down, short-range, economically-based hack and cut approach. Because economics is the driver, we simply have to decrease the debit side of the ledger. Because of our present economic position, we cannot depend on raising the credit side. Will the impact of these economic changes go too far on the line structure and economically-based staff functions? Probably. But consider the other option. If drastic changes are not made, businesses most likely will not survive. They will never be able to posture themselves to successfully compete today nor be in a position to join those in the second phase of change and create tomorrow's businesses. Wisdom, of course, could be used to determine the exact level that is required without dissecting the line or staff too deeply. But remember, we have never been here before. Therefore, the wisdom doesn't exist. We will only be able to look back and analyze our successes and failures. The trick is to *be* in business tomorrow so that you can analyze the path and decisions you made.

So, will we go too far in our impact on the organization? Probably, but that problem is greatly exceeded by the problem of not going far enough, considering the goal is to have the organization survive.

Worker Needs

This is a difficult perspective because individual evaluation has two basic needs, today and tomorrow. Obviously, you must survive today to get to tomorrow. So, let's deal with today first, the phase-one impacts on the worker. Realistically, workers have many other needs from work beside safety and health, including pay, job security, a feeling of self worth, and a feeling of organizational worth. Phase I changes cannot help but step on a number of these worker needs. But things cannot remain unchanged for the worker if business and the organization must be changed. The worker and his or her needs are going to be impacted. Basically, there will be changes in jobs. Some will continue employment and some will not. It will be based a lot on skill or technical requirements of the emerging workplace and what "pain" the particular business will have to go through to survive. Most assuredly, no one's job will remain the same. Job security will also change. Economists tell us that in tomorrow's work world there will be a lot more job hopping, temporary work, and self-employment. Each of these speak to changes in job security. Economists also tell us that we will be making less income in the future. We've seen this change coming for the last twenty years or so. That's why two-wage-earner families are the norm today.

Areas of impact including feelings of self worth and organizational worth, and safety and health are the real unknowns. They will also be highly organization-to-organization dependent. Traditional American business ways have held to a devaluation of these basic needs. Is this going to change in the short term? Probably not. Personally, I think it will get worse before it gets better.

Individually, will these changes go too far? Not in businesses and organizations that currently have worker needs as cultural values. These are cornerstones to these businesses. But these businesses and organizations are the minority, not the majority. As change impacts these organizations, their value system will remain intact. But what about those businesses and organizations that presently do not value worker needs? This is the ugly side of the picture. Economically based decisions will only make things worse. Because of this, they probably won't survive into phase II.

What about the second phase changes? How will they impact workers? Let me answer this with another question: What is the range of different ways workers are valued in today's workplaces? If we focus on this question, knowing that the answer is "very wide, from no value to high value," we can find solace in tomorrow's workplace. Tomorrow's workplace will be based on greater worker participation in all aspects of the business. It can be no other way. If this is the case, the value of the worker to that business will logically be higher than it is today. In tomorrow's workplace, the "me management, you labor" paradigm cannot exist. This is good news for workers, but not necessarily for unions as they define themselves today.

The Changing Practice of Safety and Health

If we focus only on the traditional practice of safety and health and are not flexible enough to find our new role in tomorrow's workplace, the news is not good. Changes in America's businesses and organizations will impact our turf. It will be fundamental in its assault. Our practice will change and it will change radically from our traditional thoughts. Already we see the changes occurring. Safety and health staffs are being right sized. Services are being out sourced.

A LOT OF GROUND TO MAKE UP

American industry has a lot of ground to make up. We not only waited until the competition had passed us, we waited until they were well ahead. In a large way, this is thanks to the protective mechanisms we placed in our competitors' way like tariffs and import restrictions. Governmental actions requested by business built a protective moat around our sagging business practices.

Let me use an example to illustrate the amount of ground we need to make up. Compare the time required in the automobile manufacturing industries in both America and Japan. Look at three key cycle times[5]: sales to distribution, vehicle manufacturing, and new vehicle introduction.

---

[5] Cycle time is the amount of time it takes to complete a task from start to finish.

The cycle time for sales, ordering, and delivery of a special order vehicle to a customer in America is between 16 and 26 days. In other words, the time between when the sales agreement is completed at the dealership and the car is delivered for customer pick up ranges from 16 to 26 days for an American automobile manufacturer. In Japan, this same cycle takes six to eight days. Which customer, the American or the Japanese, will be more pleased with his or her special purchase? Where will customer satisfaction be the greatest?

The cycle time for vehicle manufacturing, from start to completion of an automobile, takes 14 to 30 days in an American car factory. In Japan, it takes between two and four days. Assuming equal pay rates for workers, which production line costs less money to manufacture a car? Which manufacturer can afford to give a better deal to the customer and still make a healthy profit?

How much time does it take to take a new vehicle from concept, to design, manufacture, and introduction to the consumer? In America, that cycle time for a new concept takes between four and six years. This is a radical improvement over the amount of time it took us just 10 years ago. In Japan, however, this same cycle takes only two-and-one-half to three years. Which car is going to be the most technologically advanced when it hits the showroom floor? Which new model will have technology that is two to three years behind what the other country's car manufacturer is providing for sale? The simple fact is that due to these cycle times, Japanese car manufacturers can please customers more, sell for less and sooner, and provide a new design that has a median technology age of only three years compared to a median age for American cars of five years.

Now ask yourself—does American industry have a lot of ground to make up?

There is an analytical tool that is used in business circles to tell which businesses or segments of a business are valuable contributors. It is called the growth-share matrix. The matrix is divided into four quadrants of a square. The vertical axis represents the growth potential of the market that the business or segment works in, low growth to high growth potential.

The horizontal axis represents the competitive position that business or segment holds, high competitive edge to low competitive edge. The matrix looks like this:

High Growth **High Competitive Advantage**	**High Growth** **Low Competitive Advantage**
Low Growth **High Competitive Advantage**	**Low Growth** **Low Competitive Advantage**

Each box within the matrix represents the viability of a business or a segment of a business. For example, businesses that have both high market growth and high competitive advantage and occupy the upper left box are self-sufficient in cash flow. These are rapidly growing businesses that are large net generators of cash. They are called "Cash Stars." This is the box where American business found itself following World War II and continued as other countries developed. In those times we were a heavily exporting country. We were rich and our businesses were "Cash Stars."

As the business sectors of other countries grew, American business moved into the lower left box. Here, businesses generate far more cash than they can profitably reinvest. These businesses are called "Cash Cows." As the businesses in other countries became highly competitive, American business moved to the lower right box. Businesses in this box generate far less cash, however, theoretically, their demand for cash is also smaller. These businesses are called "Cash Traps" because they reinvest every penny they have, especially if they have a tradition of doing so, like American business did. In other words, they continued to spend and began spending more than they had.

As American business continued to deny that global business had changed and that their market and advantage had eroded, the technology revolution happened. This drove the growth curve for businesses back into high gear. There was one problem: American business hadn't conserved their cash nor had they made the changes to ready themselves for such a market change. They were now in the upper right box where demands on their businesses demanded far more cash than they could generate. Businesses in this box are called "Question Marks" because they are in a highly vulnerable economic and market position and because of that, they may not survive.

Being a "Question Mark" in the global economy places severe demands on businesses to change or fail. This was exactly the cause of the first phase of business change—they got to a position where cash was stabilized and available.

ARE THE CHANGES GOOD OR BAD?

The evolution of American business impacts worker safety and health. There is no question about this. But the changes that American business is now going through are necessary if we are going to remain economically independent and healthy as a nation. Knowing that these changes are necessary, good or bad, we need to focus on the impacts of those changes and things that stand in the way.

Fact: American business is changing. The first phase is economic. It focuses on positioning for competition. The result of phase one (for surviving businesses) will be leaner and flatter organizations. They will have to do things differently and more efficiently, or not do them at all. Until organizational streamlining can occur, however, there will be a demand to do more with a lot less. Worker safety and health will be guided during this phase by the traditional values of safety and health in each business. Those companies that traditionally hold worker safety and health at high value levels will continue that belief. The impact on safety and health in those businesses will be minimal. Those businesses that have traditionally held to low value or no value for worker safety and health will probably degrade those values further. This will occur totally because of the value system that is in place. Secondary values or priorities get lost or forgotten when primary values are at risk and need constant attention. In these companies, phase one will bring worse levels of worker safety and health. This further devaluing of worker safety and health will heavily impact the ability of those companies to detain necessary participation and buy in to change. This will place these companies at a significantly higher risk of failure.

The second phase of American business change will have the survivors of phase one reinventing American business. In phase II, worker safety and health will be better across the board because of the need for high participation and ownership in tomorrow's successful workplace.

The professions in safety and health, however, will be vastly different from how we have traditionally known them. There will be less defined turf for safety and health officers. It will be more of a grassroots, shared responsibility throughout the line structure. Specialty services that are required will either be done by support staff or by outside resources on an as-needed basis.

ROADBLOCKS TO CHANGE

What roadblocks stand in the way of this necessary change in American business? I can see three significant potential roadblocks: traditional short-range thinking of American management, labor contracts that interfere with change, and safety and health professionals not embracing change.

Short-Range Thinking

The first and most significant potential roadblock is the traditional short-range thinking of American management. You simply cannot go into change, especially phase one, with a three-month meat cleaver approach.

The shock of becoming anorexic from cutting away too much meat with the fat could be a disastrous handicap to any business. Being forced to grow and regain muscle when you should be systematically reducing can create havoc and mission confusion. Those companies that enter phase one and stay thinking "the next quarter's financials," instead of focusing on a long-range vision of where they need to go, will likely not survive this phase. At least the management team won't.

Labor Contracts

The second potential roadblock is labor contracts that impede or greatly slow down change. Change must occur. There must be more active participation, more flexibility between crafts and trades, less dependence on acquired "turf" and protections. Tomorrow's working world can be either different or nonexistent. If we accept the reality that it will be different, labor can be a power in helping shape that new working world. Holding on too tightly to what has been acquired in the past could very well be the death knell to whatever company or business sector labor moves against. This change depends on participation. Taking sides and resisting change will have a negative impact on phase one. But this is a two-way street. Management must not return to the "me management . . . you labor" philosophy or they invite non-participation in the change process.

Refusal to Change

The third potential roadblock is safety and health professionals not embracing change. Worker safety and health will have to be done differently in the future. As professions, we have a choice, run toward the light or drag our feet. If we choose to participate in the creation of this new mechanism for assuring worker safety and health, we will evolve as professionals and as professions. However, if we choose to drag our feet on change as we have traditionally done, we will doom ourselves and our professions. There will be no place for non-team players in tomorrow's workplace. Without jobs, the profession will shrink, or worse. This, of course, is our choice, individually and as professions. We had better start preparing, one way or the other.

SUMMARY

Current problems—what we should stop doing:

- Thinking that the change occurring in American business is a fad and not necessary.
- Thinking that the only acceptable change is one that does not impact us or take any turf away.
- Thinking that we can sit idly by and let the change occur without our involvement.
- Thinking that our world will be the same during and following these changes.

Fixing the system—what we should be doing:

- Getting actively involved in the changes to American business.
- Preparing ourselves to be valuable contributors in tomorrow's workplace.
- Helping change so that worker safety and health issues are minimally impacted during change.
- Learning as much as we can about change and change methods.

9

WHERE DO WE GO FROM HERE?

We tend to be too focused on single issues. This, of course, is an American trait. We see it so often. For example, how many politicians get elected based on their position on a single issue such as crime or balancing the budget? But doesn't an elected official's job have a lot of different requirements and skills, and aren't there many important issues they will need to address? We tend to forget this when we elect them. However, when the elected official doesn't show up at work or at public meetings, is a crook, can't balance his own check book, or has such terrible management skills that he nearly causes his staff, the police department, to walk off the job, we just vote in someone else. We move on to the next single issue.

This is the same way we deal with worker safety and health in this country. We've turned it into a long string of single issues. When it was a financial disruption to business, we started workers' compensation. When there wasn't enough clout for worker safety and health, we created OSHA. When safety and health guidelines were not automatically implemented in industry, we made them law. When we couldn't get management to take safety and health seriously, we made it a part of our labor contract and required a Joint Union and Management Safety Committee. When the professions of safety or health weren't getting enough respect, we grew our professional organizations so that no one could ignore us. More often than not, this approach at problem-solving has focused on the symptoms or inconveniences, and not on the problems at all. We call this approach "shotgunning." Rather than taking aim at a particular target and hitting it, the shotgun approach hits everything and nothing at the same time.

A manufacturing company's products had a cosmetic defect appear on their surface. Did it impact the product's performance? It didn't matter. The customer didn't like it. The company became very concerned. That product represented a sizeable percentage of their business. How did they resolve the pressing concern? Well, a logical and meticulous person would have carefully investigated the problem until he found the cause. Then, he would have worked quickly to remove the problem. The rub with this approach is that most often it takes a lot of time to isolate and identify the problem. Remember—we are Americans so patience is not a traditional characteristic. So, with the company management being impatient and very worried about pleasing the customer, what did the company do? They quickly changed a number of things to see if the defect would go away. It did. Triumphantly, they announced victory to the customer. The customer wanted to know what they did to correct the defect. Not knowing which action they took was actually the solution, they told the customer *all* the actions they took. "Good," the customer said, "make all my products with those added steps included." This, of course, added production costs to the company and resulted in no increase in selling price. Net result—less profit and more work. One year later, the surface defect showed up again.

Why is this story significant to our discussion? There are two important points that we can learn from this example. First, throwing solutions at a symptom gets you nothing. Actually, you can't solve a symptom at all. You have to know what the problem is and solve it. And second, we need to develop a little patience in our problem-solving, realizing that spending a little extra time isolating and identifying the problem saves us a lot down the road. Is the level of worker safety and health in America today better or worse than in the past? The truth is we don't know. Industry has changed. The way we measure things has changed. But really, isn't the answer to this question fundamental to our mission? How can we have continually added "things" to the system for worker safety and health, focusing on single issues only, and not ever know if worker safety and health was getting better or worse? Our history tells us that we've been shooting in the dark at symptoms with a shotgun.

A "DEBRIS" SUMMARY

Let's stand back and take a look at the entire forest, the system for worker safety and health. Seeing the forest, we can identify the various trees, the parts that make up the system. We've included six parts in this analysis. They represent the major parts of the system for worker safety and health in America. These include the workplace (which can be divided into the organizational structure and the business itself), the regulatory environment and the legal requirements, the workers' compensation programs, organized labor, the liability issues, and the professions of safety and occupational health. Each part of the system has unique origins, complexities and issues. Each has positive aspects and definitely negative aspects. It is in our understanding of each part's positive aspects and negative aspects that we can get a clearer image of the system for worker safety and health and through that image, see how we can improve the system.

The Workplace

Fundamental to this issue of worker safety and health is the workplace. From the very origins of the safety movement, during the industrial revolution, there have existed traditional impediments to worker safety and health that exist, to an even greater extent, today. These significant impediments include the economic-basis for safety, the alignment and "burial" of safety and health efforts in the power-struggle of the staff organization, the lack of recognition that safety and health was a line responsibility all along, and the building and nurturing of a caste system between management and labor. Are these impediments chiseled in stone, never to be changed? One would think so, looking at how deeply they have become embedded in our American business culture. From workplace to workplace, it's almost a cookie-cutter image. No wonder labor organized, and that labor laws and inspections became acceptable solutions to this maze.

A separate but highly influential aspect of the workplace are the changes that business is and will be, undertaking to survive in the global economy. It's a shame that we can't deal with improvement of worker safety and health issues within the organizational structure before getting into real change. But, everything seems to happen at the same time. At least that's the way it is happening now. Make no mistake about it—American business is changing. It's a must in order to survive into the next century. But change isn't just happening to business. Everything associated with our economy will change. This, to the dismay of those that have chosen the "safer" ports, will impact public service, academia, and government. These have to change also. At their present level of inefficiency and ineffectiveness, they place too great a burden on our entire economy. As businesses must change, our total economy must change also.

This, of course, is good and bad. It is good for business survival and prosperity. It is good for our economy. It is good for continued employment for us, our children and their kids. It is good for our nation and our ideologies. And, because of what America can contribute, it's good for the world and the world economy. But it also has a bad side. Things will and must change. With that change, however, many aspects of our lives will also change. These include our job security as safety and health practitioners, organized labor as we know it, our nation's per capita earning power, the kinds of jobs that will be available in tomorrow's workplace and some aspects of our society. These, in comparison to today, will be negatives.

Worker safety and health should be a positive, though. With the high participation required in tomorrow's workplace, worker issues such as safety and health should enjoy a revival. Since this is our mission, it is a very positive result of the changes in American business and the waves it will cause throughout our corporate structure and economy.

Regulatory Environment and the Regulations

All aspects of the regulatory environment for worker safety and health grew out of the problems or ineffectiveness of worker safety and health in the workplace. If that hadn't been a problem and worker safety and health had been a value, a cornerstone of American business, the regulatory environment wouldn't have thought of, much less created. But it wasn't and, therefore, the regulatory environment and the regulations were created in an effort to plug a gushing hole in the dike. Did it accomplish that? Remember the example of the company that made the metal product that one day had a surface defect? Just as that company in a panic threw ineffective solutions at the symptom, we threw OSHA and the regulations at the symptoms of poor worker safety and health. "Something has to get industry's attention" was the battle cry. Today, well over twenty years later, when the existence or mission of OSHA is threatened by Congress, you hear the very same arguments.

Are the regulatory environment and the regulations what we originally intended them to become? No, they are cumbersome, ineffective, inefficient, can't keep up with new technology, can't inspect all the workplaces, are stupid, too political, change in direction with the wind or changes in the "ruling party" (same thing), and pretty well have been a bust since the opening volley (good analogy when you think of shotgunning).

I recently read a couple of letters that were printed in the "Letters to the Editor" of a national publication. One stated an emotional position concerning what would happen to the jobs of safety and health practitioners should the teeth be taken out of OSHA. The other stated that only OSHA could make business safe and healthy. Haven't we really dispelled these ways of thinking? First, the changes in business are going to change our jobs anyway. What possible impact could OSHA's having more or fewer teeth cause in this changing business environment? If at all, it could only affect the change in the short run. Second, has the "toothed" OSHA been effective at "making" business think differently about worker safety and health issues? Maybe it has had some minor impact on the way that some business leaders think about compliance and the risks that go with it, but it has had little or no impact on changing the way our business culture thinks about worker safety and health. The only possible origin for a change of that magnitude is from within business.

Workers' Compensation

The workers' compensation programs were a direct result of a workplace system for safety and health that did not work. In reality, it was a creation by both government and business to control costs. It's a shame that safety and health got a toehold in America as an economic issue—a cost containment or avoidance strategy. Unable to shed this attitude, we've accepted it in industry, in government, and in labor. It has become chiseled into the granite of our culture.

Because workers' compensation was created as a special program focusing on cost containment, it has suffered many predictable and unavoidable problems. There is not one program for workers' compensation, there are over fifty of them. The basic structures are similar but the details that directly impact its ability to balance cost effectiveness and ease of service vary greatly. This not only presents significant problems to multi-state companies but great challenges to state economies that try to attract business expansion and new businesses.

The workers' compensation program is an established and controlling monopoly. Monopolies can have tremendous problems. Efficiency in monopolies is rare and motivation is even rarer. These drive up the costs to business and society. Monopolies also tend to control the market. Workers' compensation programs are no exception to this rule. Ask any independent insurance company. Monopolies in control of the market expand uncontrollably, often at the disadvantage of private industry. The safety departments of state workers' compensation programs are good examples of this expansion. This expansion not only drives up costs, it limits expansion of the private sector that pays a significant portion of the tax that society uses.

Workers' compensation has its own highly political court system. Often this system defies our founding fathers' vision of blind justice. Politics and justice too often lie at opposite ends of the spectrum.

The largest problem with workers' compensation, like most specialized programs, is that it works from exclusionary concepts. Its heavy administrative burden and application of benefits lie totally on work-related cases. This was originally justified from a "charge back to business" cost basis. In today's world, where there are competing systems for health care, this rationalization is a dinosaur. Programs based on exclusionary concepts have tremendous problems and heavy costs.

Organized Labor

This is not a sacred cow. Like the other parts of the system for worker safety and health, organized labor has had both positive and negative impacts. Admittedly, considerably fewer than half of America's workers are organized today. But the impact of organized labor extends far beyond its membership into almost every workplace. It establishes the measuring stick for many worker issues, including safety and health.

The problems include those that occur when working by contract. Contracts are based on distrust, the belief that the only way the other party will do what they say is to tie them to a written agreement. It is far from the old-fashioned handshake. Contracts are written guarantees, but more importantly, they are limiting instruments. They establish boundaries. Boundaries not only place limitations upon participation, on both sides, but they make efforts within those boundaries almost impotent.

The designated union hierarchy and established communication pathways also get in the way of timely, effective, correct and non-political communication. After all, if the safety and health of workers is supposedly a high priority and requires quick resolution, having established pathways only flies in the face of accomplishing this. In most cases, it is a significant roadblock.

Power-based negotiation and bartering tactics are impossible to avoid in a contractual relationship. That's where politics comes in full steam and power becomes all important. This is a no-win situation for workers because they will never have the power that management has and countering that power with politics only confounds their softer issues like safety and health.

Contracts also perpetuate the traditional management and labor roles. Traditional American business has held to the "me management, you labor" caste system. Contracts only perpetuate this stagnating tradition. Communication is filtered both ways and information becomes power. Polarization is inevitable.

Union structures are an "all for one and one for all" concept. It devalues individual issues and contributions. This is antithetical to the way tomorrow's business must be operated.

The Professions

I guess it is more or less a predictable cycle for professions. They spring up out of need, define a new profession, start grouping together for support and professional growth, grow to the point that you feel you aren't getting the respect you need, seek ways of increasing your clout by growth and absorbing other disciplines closely related to yours, reach the point where tradition is more important than change and begin a downward spiral. In short—discovery, realization, camaraderie, expansion, tradition, arrogance, confusion, and decline. It's a nine-stage process. The ninth stage is, of course, death.

The safety and health professions began from very different roots to come to almost the same place. Safety started on the factory floors practiced by laymen. Health began as a specialty in universities and research practiced by scientists. Today, however, both professions find themselves cycling between the fifth and sixth stages. At this point, it very much looks like a land grab. Good examples are ergonomics and environmental disciplines being claimed and absorbed into both the safety and health professions as theirs.

These last-half stages are where there is great confusion. The strength of what brought them to where they are is confused with where they are going. Losing roots, steeped in expanding and unmoving tradition, these last-half stages are very confusing. One of the greatest areas of confusion is in the mission—the profession and the clout it carries become more important than the original altruistic one. The professions begin to delight in themselves and forget why the professions were needed in the first place. These are ugly stages.

The Blurring Line between Legal Requirements and Liability Risks

Workers' compensation grew from the issue of liability risks. With the passage of the OSH Act, most ideological supporters felt that someday all significant or important aspects covering worker safety and health issues should have the force of law. Well, as history has taught us very well, the machinery we put in place to update or create regulations is so cumbersome and slow that it never even has a chance of catching up. This has been one of the largest sources of frustration to those who believe that the regulatory means is the only way to ensure safe and healthy workplaces.

The saying, "What goes around, comes around," seems prophetic in this issue of legal requirements and liability risks. What we created to streamline the system and avoid messy tort actions or the necessity for them seems to be coming back to center. Workers' compensation is no longer the exclusive remedy for worker injuries and illnesses. If there is negligence involved, and if you can establish knowledge, you can sue. And more and more workers are suing employers every day.

The once clear lines between legal requirements and liability risks are fading away, if they exist at all anymore. No longer can forward-thinking management take comfort and refuge in the thought that if the law didn't require it, it didn't have to be done. Factually, the costs associated with violations of law are much lower than those associated with liability. And, if liability risks are increasing at near exponential proportions (seemingly as the number of lawyers increases), it makes little sense to ignore them as highly improbable occurrences.

IF ONLY IT WERE PERFECT

I take companies and groups through a three-step process when I am guiding them through change or helping them transition into successful changes. The first step is to define, as exactly as possible, what the real, existing world looks like. What do you want or need to move away from—the starting position? This can focus on a company, a business sector, a market sector, a total economy, a program or a specific group. This is a reality check where all "truths" can be put on the table and examined. It is also a point where the group can get their frustrations out and see them for what they really are. It is a necessary part of the healing process that must be undertaken if anything is to get better.

The second and crucial step is to purge our present world reality, and try to visualize and define a world that is ideal. This part emphasizes the perfect situation, the way it could be if everything were ideal. It transforms natural roadblocks inside our minds. As we become adults, we are taught not to fantasize and to deal only with reality. We feel that dealing in what cannot be is both unhealthy and unrealistic. So we deny ourselves that adventure. This part of the change process explores the ideal, casts off our natural reluctance to do so, and allows the group to explore what an ideal world could really be. But because of our upbringing, this is not easy for us to do.

In most things, our learned reluctance is right. Thinking of the ideal is fantasy. But it is more than that. It is the substance of which dreams are made. Imagine a five-year-old that is stuck in reality. We can't. We haven't ever met one. A five-year-old is unencumbered by our adult reality constraints. A five-year-old has no problem reaching across the abyss of reality, seeing the ideal, and dreaming. In our adult understanding of reality, we call this make-believe. As we get older and more "disciplined," we lose this natural ability to dream. We simply deny ourselves this adventure. We sentence ourselves to our known reality, accepting what we are dealt. So looking at what could be, with permission of course (it's the only way we can do this), is healthy to removing our paradigms and baggage of the past and freeing our creativity. As Pablo

Picasso put it, "Every child is an artist . . . the trick is to remain an artist when you grow up." What he was talking about, of course, is our ability to dream, escape reality, and be creative.

The third part of this process is to define what barriers prevent us from realizing this ideal world. It is a way of focusing on what we can change and what we cannot. But this a subject for later on in this chapter.

If we ruled our world and could create a world including business, labor, regulations, etc., that was perfect for providing excellence in worker safety and health, what would it look like?

The Business Environment

Sometimes I envy those countries that are building economies from scratch. Because what they have is nonfunctional, or grossly inadequate, these countries move without fear of risk, almost with reckless abandon, toward change. They try anything. They learn from anyone. They change everything. The only stance that is not acceptable to them is to do nothing. Obviously, this is not the situation where American business finds itself. Here, we tend to be schizophrenic. Being trapped in our traditional business paradigms and practices, we require a major crisis to get our attention. Minor change triggers like opportunity or need have *no* chance for success. Even the higher change trigger of discomfort provokes a weak and temporary commitment to change. In such a world as this, we too often reduce our effort to that of chasing fads or the "plan of the week." When the level of discomfort decreases (most often from actions that are not ours), we quickly revert back to our traditional ways, paradigms, and fixes. It's like our driving practices when we are followed by a traffic cop. When he's behind us, our driving practices are perfect. But when the officer is no longer there, however, we quickly revert back to our traditional, less perfect way of driving.

Let's say that we could shed our traditional business paradigms and practices and instantly change to what will be the most successful. This is a stretch for even the most freethinking. But what would that business environment and those practices look like?

First of all, it would be an environment where all participants have freedom to contribute, no roadblocks to participation exist and everyone has ownership in the success of the business. Management and labor are active, communicating partners in the effort. No group is superior or inferior to the other. Everyone is equal and dependent on the other for success. It is a synergistic environment where the sum is much greater than the individual parts. It is energizing, motivating and sustaining.

Second, there would exist a dedicated effort in one long-term direction. It would also be acceptable to "expense" a little today to reach tomorrow's vision. The long-term vision would logically be more important than today's short-term thinking. Goals and objectives toward that vision would be derived not by one person but by every contributing member of the team. Planning toward that vision would weave all players, responsibilities, necessary activities and results together into the same tapestry for success.

Third, everyone would know their roles and be competent to play them—whether leaders, players, support personnel, or whatever, all would be committed to the team effort—knowing that by playing their role to perfection, the team wins. Each player would also be accountable for his or her role and responsibilities. In that empowering and accountable environment, all would have the necessary skills to play their roles, and their efforts at building tomorrow's skills would be ongoing.

Fourth, responsibility and authority would be holistic in all players, regardless of level in the organization. In such an environment, there would never be confusion about values such as worker safety and health, and responsibilities would be performed in balance. There would be no politics or responsibility-without-authority traps that sap organizational and individual strength and motivation. There would be no "my win . . . your loss," in that all team players would recognize that all players either win or lose together.

Fifth, the environment would be caring of and helpful to all individual players. There would be no confusion about the importance of the human spirit being much higher than short-term economic gain. After all, individual dignity defines the high standards and values of the organization.

The Regulatory Environment

In our perfect world, there would be no need for a regulatory environment. The regulatory environment was created solely because our traditional system was not functioning. This makes it necessary, most would argue, for a "thou shalt" component to assure that most businesses and management will indeed function. In a perfect business environment, however, there is holistic responsibility and a high value for human worth. This business environment alone makes "thou shalt" thinking unnecessary. It becomes merely the best way of doing business and treating each player. No, in a perfect world, there is no need for regulatory power. Responsibility is accepted, not required.

The Holistic System for People-Care

A billboard I saw recently said "Health coverage should not be a luxury." I couldn't agree more. But it should not be considered an entitlement, nor an individual right either. Health care should be a value that is placed so deeply within our society that it is assured to all. Of course, there needs to be responsible use and provision. You simply cannot live with a system that provides CAT scans routinely for headaches. Remember, we are considering only a perfect world. This makes this belief not only realistic, but acceptable. But there should be no exclusionary health care programs based on different causes or symptoms. Of course, there is a need to assure that costs of such a system are borne by those sectors of our society that have specific ownership. But in our perfect world, where proper health care is assured to everyone, no exclusionary provisions would exist. In such a perfect world, one health care system should take care of all medical problems. Individual compensation, if right and appropriate, and earning protection should be borne by the person's employer. Like on-the-job responsibilities that look out for the dignity of the whole person, there is no room for nine-to-five mentality in our perfect world.

Management and Labor

We are too bound by our traditional "us against them" mind set, regardless of which side of the fence we find ourselves out. In our perfect world, however, the key concept is "team." Just like the playing relationship between an offensive lineman and a running back, a defensive player and the offensive line, a wing on a hockey team and the goalkeeper, a basketball guard and a center, a defenseman in soccer and a striker, a baseball pitcher and a shortstop, there must be a team orientation between labor and management. It is not a "we win, you lose" situation. In a true and successful team, all win or lose together. In such a team environment all players are committed to the success of the team, the organization. Everyone, as group and individually, works for the success of the team. Team dynamics are based on knowledge, trust and active communication. None of these can be shortchanged or sidestepped in our perfect world. None can be withheld as a source of power, turf, or tradition.

The Professions

This effort of visualizing what our professions would look like in a perfect world should be the hardest for us. We are too close to them and many of our beliefs have been defined by the present system for worker safety and health. More exactly, however, our beliefs have been shaped by the inescapable problems within the system.

Our professions were created as a means to improve worker safety and health. It's really that plain and simple. This is a critical perspective because too often we confuse our purpose with our natural tendency to preserve what we know. Remember the mistakes made by the Swiss watch industry when they ignored quartz technology?

If our purpose is to improve worker safety and health, what is the most effective way of doing that in our perfect world? Is it via a specialized, staff-aligned function that historically has been held responsible without the direct authority to get the job done? If history has taught us anything, this is not the answer. So, where should that responsibility exist? In the line structure, of course. In our perfect world, the responsibility for worker safety and health falls solely into the line organization. What then does this perfect world hold for our profession, our speciality? If we change the way the system works, does our profession become a necessary victim to the efficiency of the system?

Not exactly. Because of the complexities and the level of knowledge necessary to do our job, there will be a need for this specialty even in our perfect world. The profession will have a specific and necessary place. Of course, the number of specialists would go down. With line responsibility, most of our day-to-day chores would be done there, not by specialists. There would simply not be a need for as many specialists as there are today our dysfunctional system for worker safety and health. But, in our perfect world, our purpose would be enhanced to accomplish what we cannot with the present system for worker safety and health.

This perfect world vision forces a face-to-face confrontation between our professional goals and our purpose as professionals. We've become very confused about the difference between these because we have not been able to transcend our present system's realities. In our present system, we have deduced that power and numbers are the only way to accomplish our purpose. In a perfect world, however, the system is totally altered. The problems are removed. In doing so, we have invalidated most of our present beliefs. If our purpose remains firm, our acceptance of the new reality should be more clear—not easier, just more clear.

The Liability Threat

I must admit I don't look at liability as a negative aspect. To me it falls into the same category as "buyer beware." One says that you must act responsibly and minimize your risk of liability. The other says that as you live, there are no guarantees. You might guess that I don't place much faith in the concept of government protection. It isn't that I necessarily dislike the thought, it is just too unrealistic and tends to add to the government's power and lessen ours. I don't think our country's forefathers had that in mind and I'm not willing to give up some of my individual rights or options in exchange for bureaucratic oversight. Frankly, it is one of the burdens that brings this country down, not adds to its strength.

The role of liability is constitutional in origin. It is basic law. If you don't act responsibly, you might end up paying handsomely for it. We understand why the remedies for worker safety and health problems have been designed the way they are. It creates a no-risk scenario for both business and individual workers. But that it is one of the major reasons why we have the paradigms that created the regulatory process. Perhaps it is time that we return to that constitutionally-based role. Removal of the regulatory "thou shalts" and the automatic safeguards of the workers' compensation programs would leave us with only tort law relief. Is that bad? If we believe that the only "hammer" that gets business to pay any attention to worker safety and health issues is the law, why are we still placing our faith in a regulatory process that can never catch up? In a perfect world for worker safety and health, where both worker and business are committed to improvement, liability risk is the best pathway.

Why? First, it is not dependable nor is it automatic. Workers must go to bat for what they believe and business must protect itself from risk. Second, without safeguards, the costs are much greater. OSHA fines cost industry little, if anything, other than bad publicity. Torts do more. Sure, they are more messy and require more time, but perhaps with the abundance of legal services we have today, it is indeed time to change.

Throw away the regulations. Bring on the suits.

ROADBLOCKS TO EFFICIENCY

This is the third part of the change process I share with clients. Once we have defined the real world, and have visualized what it could be like in the perfect world, all that remains is to talk about what stands in our way of attaining this perfect world.

I love how children at play so easily shift gears. One minute they are Captain America, the next a police officer, then an astronaut. They are able to make drastic changes in roles and situations with ease. You might argue that they only play mind games and that reality is very different. You're right of course, but from the mind comes our ability to change in the first place. If we cannot regain that ability to make mental changes, we can never begin the process of changing reality.

But mentally changing only prepares the field. At this point, massive reality-based roadblocks spring up to block our path. These are predictable. They are based in the world we have created, learned to survive in and, in a lot of cases, learned to be successful in. So we are "attached" to this reality, no matter how dysfunctional and inefficient. Reality becomes part of our predictability, which is a critical need for us to maintain. But if we are indeed trying to improve the system (our reality) so that our purpose can be truly realized, we need to look at natural roadblocks to change. They usually fall into five areas: turf, politics, fear, careers, and the effort that is necessary.

Turf

In most ways, this is a security issue. We define our security by the territory we control. Just like a wolf that "marks" his territory as a means of security, we define our turf as our assigned responsibilities. More extensively, turf includes office (and square feet of office space), staff, position in the organization (reporting relationships), lab space, equipment, etc. But turf issues can become more significant than this basic security need. In areas where there is a loose association with value (or money you bring in), like safety and health, turf can become a significant part of one's power base. The greater the turf, the greater the power within the organization.

These two turf issues, security and power, are, of course, very different. Together they provide a full spectrum of possible roadblocks to change. Maslow tells us of the basic, primal need for security. Power, however, is a higher-order need. It is an ego issue. Because of these differences, a perception of threat to turf can be met with very different reactions, based on area of need. From a security viewpoint, there is give and take. There is no primal quantification of need, only that something fill the void. If there is turf (no matter how large or small), there is security. Power issues are very different. They are definitely quantifiable. The bigger the slice of pie, the better. But any perceived threat of reducing the amount of turf is met with aggressive resistance. In a power issue, the size has a direct relationship with need. The less the turf, the less it fulfills the ego.

It is natural for those of us who do not have direct responsibility for our organization's earnings to deal with turf as more than a security issue. It becomes an ego issue. Look at how important titles and credentials are to us. The title "manager" is much more impressive than "engineer" or "industrial hygienist" and the title "director" is even greater. Look at our business cards to see how important credentials are to our ego. Very seldom do you find a certified professional, especially if he or she has an advanced degree, that doesn't include those treasured letters on his or her business card. It is part of our turf and it helps feed our ego.

Turf, therefore, can become a significant roadblock to change if that change has a perception of less importance or fewer jobs. We resist the thought because our dependence on turf to feed our security and ego needs is strong.

Politics

Webster defines politics as being shrewdly tactful. The word comes from the Greek root *politikos*, meaning citizen police. Where Webster got "shrewdly tactful" from its root meaning, a cop without a uniform, I will never know. In any event, as we use the word today, it is the ability to use one's power and influence to get what one wants. Political is not necessarily a friendly nor admirable adjective today. After all, most opinion polls place politicians (those whose job it seems, is politics) at the bottom of the respect scale, somewhere near used car salesmen and convicted felons. Why would politics then be a roadblock to change? Have you read anything about what is going on in Washington, DC, lately? How would you say normal politics treats anything that is different or constitutes a radical change? Politics, as we know it today, tends to maintain the status quo and resist change. Phrases like, "If it ain't broke, don't fix it" or "It ain't wise to send a fox to count the chickens" come to mind (spoken in a Southern drawl, of course).

The key point is that politics uses one's power base to produce a stalemate. And there are many key players to change in the system for worker safety and health that have a considerable power base. Players include business, management, workers' compensation programs, elected legislators, appointed bureaucrats, administrative law judges, union hierarchy, etc. Making sweeping changes to improve worker safety and health impacts all of these players. So politics is a very impressive high-level roadblock to change.

Fear

Don't disregard this as a significant roadblock to change. First, fear is a response to change. Second, there are many faces of fear. Some of the greatest include fear of the unknown, fear of not being needed anymore, fear of incompetence, fear of losing turf or power, fear of more work, fear of losing control, fear of someone else getting credit, fear of learning new skills, fear of not being as sharp as you used to be, fear of not being able to adapt, fear that the outcome will not be better, etc.

Most of these come from a very basic and powerful fear—the fear of losing predictability. We are patterned animals. We like things to be predictable in our lives. Think about it. What would be your response if suddenly everyone started driving on the left side of the street or stopped at green lights? What if you came home from work and found a new child and different furniture in your house? What is your response when your spouse borrows your car, without telling you and you find it missing from the space where you left it? What do you feel when you reach into your pocket for your car keys and find them in the ignition of your locked car? Why do you think spouses and children tend to stay in abusive homes and relationships? We have a great fear of losing predictability. Here's the rub. Even if what we have is abusive, doesn't work (like our system for worker safety and health), or is just plain bad, we tend to resist change because we fear loss of predictability.

Career Protection

Each of us has contributed a lot of energy, preparation, and time to our careers. Along comes some nut who tells us that in order to improve things, we need to place our careers in jeopardy. For any of us to place our careers at risk, we had better be very convinced that the change is better, much better that what we have today. Buying into change requires both faith and the prospect of personal gain. Even if you believe with all your heart that sweeping changes to the system for worker safety and health is best and will work great, that faith alone will not suffice. Knowing that you may gain nothing personally derails the change train. You simply don't climb aboard.

The Effort That Is Necessary

Let's discuss the positive. If we have the courage and see the job through, massive changes to the system will vastly improve worker safety and health. This will accomplish the main purpose of our professions. Now, here's the negative: In these changes there will be a lot less need for our jobs, including bureaucrats, compliance inspectors, practicing professionals, consultants, workers' compensation employees, and more. There will probably be a need for more attorneys. Our professional organizations will have to be totally revamped, redirected and will have many fewer members. Our jobs as we traditionally practice them would be vastly different. And it would take a lot of work by all of us to bring about this quantum of change. Looking at this change in this light, can you see why the effort that is necessary is a significant roadblock?

In short, there is no free lunch. What your Dad told you when you were young is correct: You cannot get something from nothing. Even if we choose to sit on the sidelines and let change happen, all of us will pay. We cannot escape paying if we are to really accomplish our professional purpose—to improve worker safety and health. The question then becomes, are we truly committed to our purpose or are we more committed to predictability and the status quo?

WHERE DO WE GO FROM HERE?

I have a confession to make. The purpose of this book was not to provide all the answers for improving worker safety and health in America. Realistically, there are more things to consider than what is included here and the possible answers are as many as the grains of sand on a beach. You may have found some thoughts that you agreed with. You may have found some discussions that you didn't. That was the real purpose—to make you think.

We are not chained to the system that we have. The system was not created by divine intervention, nor is it blessed by any higher authority. We are not genetically programmed to do things the way we do them. Like Darwin's fish that crawled out of the sea onto dry land, the system has evolved into what it is today. In reality, no one guided its development. It merely evolved the way that different species do—in response to certain environmental stressors, certain mutations are more successful than others and they survive. We also have no Endangered Species Act for the system that we've created either.

From its roots in the ineffective workplace, the system for worker safety and health grew. As it grew, it appeared to be getting better. The theory is that as you continue to improve something, it gets better and better. But the environment changed also. The parts and programs of the system became too big, lost focus, became ineffective and inefficient, and lost touch with the changing environment. What we have today is like a whale with legs on land, unable to walk or feed itself, sustained only by our continual efforts, but still expected to win races. It's time we quit feeding the creature and let it meet its evolutionary end.

We must choose what master we serve. Is our purpose to improve worker safety and health or is it to maintain the system and the predictability it provides? A story is told about a traveler in the 1800s who, at great risk and diligence, traveled across the continent from Europe to Tibet. He was seeking an answer to the riddle of life. He believed that only the Dalai Lama could provide the answer. So he traveled, risking weather, thieves, drought, and starvation to find the Lama. After years of traveling he found him high in the Himalayan mountains. Sitting at the feet of the Dalai Lama he asked, "Master, I have traveled far to learn the meaning of life. Please tell me what I could be." The Lama looked down at the traveler and stoically said, "Know thyself."

If we are committed to change, there are many challenges ahead and all of us must work for those changes. Like that traveler, there will be considerable risks and it will take time to traverse that canyon of change. But in this effort, we must know ourselves first. Then we must be committed to our purpose.

As I stated earlier, I happen to be an optimist. I'm sure that it isn't anything genetic and it probably wasn't the result of upbringing. My optimism probably came from a handful of very significant people I have been lucky enough to know at various points in my life. I also happen to be a risk-taker. I drove formula race cars as a teenager (I said nothing about being smart). But being an optimist and a risk-taker has some real advantages. I have great faith that the future is going to be better, not just a little but a whole lot. And I don't mind stepping out there on faith to try the water (or air). We are going to need a lot of optimistic risk-takers to get this job done!

Knowing the inefficiency and ineffectiveness of what we have today, the advantages of changing the system for worker safety and health are obvious. I believe that America should have the highest level of worker safety and health in the world. Our workers deserve nothing less.

BIBLIOGRAPHY

Albrecht, Karl, *The Northbound Train*, American Management Association, New York, NY, 1994.

American Society of Safety Engineers, *The Safety Profession, Year 2000*, American Society of Safety Engineers, Des Plains, IL, 1993.

Anderson, Ronald, A., *Business Law*, Southwestern Publishing, Cincinnati, OH, 1980.

Clayton, George D. and Clayton, Florence E., Ed., *Patty's Industrial Hygiene and Toxicology*, Volume I, Part A, John Wiley & Sons, New York, NY, 1991.

Coonradt, Charles A., *The Game of Work*, Shadow Mountain, Salt Lake City, UT, 1984.

Drucker, Peter F., *The Frontiers of Management*, Harper & Row, New York, NY, 1986.

Drucker, Peter F., *The Practice of Mangement*, Harper & Row, New York, NY, 1986.

English, William, *Strategies for Effective Workers' Compensation Cost Control*, American Society of Safety Engineers, Des Plains, IL, 1992.

Federal Register, Volume 37, Number 202, Part II, Department of Labor Ocupational Safety and Health Administration, *Occupational Safety and Health Standards*, Wednesday, October 18, 1972.

225

Hadden, Susan G., *A Citizen's Right to Know—Risk Communication & Public Policy*, Westview Press, Boulder, CO, 1988.

Henderson, Bruce, D., *The Logic of Business Strategy*, ABT/Ballinger Publications, Cambridge, MA, 1984.

Howard, Philip K., *The Death of Common Sense*, Random House, New York, NY, 1994.

Japan Human Relations Association, Ed., *The Improvement Engine*, Productivity Press, Portland, OR, 1995.

Larkin, T.J. and Larkin, Sandar, *Communicting Change—Winning Employee Support for New Business Goals*, McGraw-Hill, New York, NY, 1994.

Manuell, Fred A., *On the Practice of Safety*, Van Nostrand Reinhold, New York, NY, 1993.

McNally, David, *Even Eagles Need a Push*, Dell Trade, New York, NY, 1990.

Moser, Royce, Jr., *Effective Mangement of Occupational and Environmental Health and Safety Programs*, OEM Press, Boston, MA, 1992.

Nair, Keshavan, *A Higher Standard of Leadership—Lessons from the Life of Gandhi*, Berrett-Koehler, San Francisco, CA, 1994.

National Institute for Occupational Safety and Health, *The Right to Know: Practical Problems and Policy Issues Arising from Exposures to Hazardous Chemical and Physical Agents in the Workplace*, Cincinnati, OH, 1977.

National Institute for Occupational Safety and Health, *The Industrial Environment—It's Evaluation and Control*, U.S. Government Printing Office, Washington, D.C., 1973.

Ohno, Taiichi, *Workplace Management*, Productivity Press, Cambridge, MA, 1988.

Pierce, F. David, *Total Quality for Safety and Health Professionals*, Government Institutes, Rockville, MD, 1995.

Platt, Suzy, Ed., *Respectfully Quoted*, Congressional Quarterly, Washington, DC, 1992.

Pritchett, Price, *New Work Habits for a Radically Changing World*, Pritchett and Associates, Dallas, TX, 1994.

Public Law 99-499, Emergency Planning and Community Right-to-Know Act, October 17, 1986.

Public Law 91-596, Occupational Safety and Health Act, 91st Congress, S. 2193, December 29, 1970.

Ramazzini, B., *De Morbis Artificum Diatriba*, 1700 and 1713.

Sanno Management Development Research Center, *Vision Management - Translating Strategy into Action*, Productivity Press, Portland, OR, 1992.

Scherer, John and Shook, Larry, *Work and the Human Spirit*, John Scherer & Associates, Spokane, WA, 1993.

Shiba, Shoji, Graham, Alan and Walden, David, *A New American TQM - Four Practical Revolutions in Management*, Productivity Press, Portland, OR, 1993.

Stalk, George, Jr., and Hout, Thomas M., *Competing Against Time*, The Free Press, New York, NY, 1990.

Stuckey, M.M., *Demass—Transforming the Dinosaur Corporation*, Productivity Press, Cambridge, MA, 1993.

Thomas, H. Greenwood, *Safety, Work, and Life—An International View*, American Society of Safety Engineers, Des Plains, IL, 1993.

Thompson, LeRoy, Jr., *Mastering the Challenges of Change*, American Management Association, New York, NY, 1994.

Wolff, Kenneth M., *Understanding Workers' Compensation*, Government Institutes, Rockville, MD, 1995.

INDEX

A

Administrative Law Judges
(ALJ), 80, 141
Allegheny County death
calendar, 4
American Academy of Industrial
Hygiene (AAIH), 165
American business and
quality, 39
American Conference of
Governmental Industrial
Hygienists (ACGIH), 10, 14,
129, 165
American Gas Association
(AGA), 130
American Industrial Hygiene
Association (AIHA), 164
American and Japanese business,
example, 196
American National Standards
Institute (ANSI), 10, 14, 130
ANSI standards, 118
American Petroleum Institute
(API), 130
American Society of Manufacturing
Engineers (ASME), 130
American Society of Safety
Engineers (ASSE), 164
aspects of business, 21

B

balance sheet, 186
Barker, Joel, 42
bricklayers, example, xv
bottom line, 21, 131
*Brown versus Long Island
Railway*, 6
business
and change, 42
and power, 180
and quality, 39
change cycle, 185
consumer focus, 182
creation of consumer need, 183
downsizing, 186
economics and safety and
health, 7
environment, solutions, 213
expansion and contraction, 185
fad chasing history, 187
flatter structures, 189
growth-share matrix, 197
labor contracts, 200
less supporting staff, 122, 191
management style, 181
objectives, 39
patience, 145
polarization of management and
labor, 181
priorities, 39

Environmental and Health/Safety References

Total Quality for Safety and Health Professionals

F. David Pierce, a CSP and a CIH, shows you how to apply concepts - proven successful - to your safety management program to achieve increased productivity, lowered costs, reduced inventories, improved quality, increased profits, and raised employee morale.
Hardcover, 244 pages, June '95, ISBN: 0-86587-462-X **$59**

OSHA Technical Manual, 4th Edition

This inspection manual is used nationwide by the U.S. Occupational Safety and Health Administration's inspectors in checking industry compliance with OSHA requirements. *Softcover, 400 pages, Feb '96, ISBN: 0-86587-511-1* **$85**

Exposure Factors Handbook, Review Draft

The U.S. Environmental Protection Agency uses this document to develop pesticide tolerance levels, assess industrial chemical risks, and to undertake Superfund site assessments and drinking water health assessments.
Softcover, 866 pages, Nov '95, ISBN: 0-86587-509-X **$125**

Ergonomic Problems in the Workplace: A Guide to Effective Management

The valuable insights you'll gain from this new book will help you develop and implement your own successful ergonomics program.
Softcover, 256 pages, Sept '95, ISBN: 0-86587-474-3 **$59**

Product Side of Pollution Prevention: Evaluating the Potential for Safe Substitutes

This report focuses on safe substitutes for products that contain or use toxic chemicals in their manufacturing process.
Softcover, 240 pages, Sept '95, ISBN: 0-86587-479-4 **$69**

"So You're the Safety Director!" *An Introduction to Loss Control and Safety Management*

This book concentrates on your role in evaluating, managing, and controlling your company's losses and handling the OSHA compliance process.
Softcover, Index, 186 pages, Oct '95, ISBN: 0-86587-481-6 **$45**

Emergency Planning & Management: *Ensuring Your Company's Survival in the Event of a Disaster*

This book will help you assess your exposure to disasters and prepare emergency response, preparedness, and recovery plans for your facilities, both to comply with OSHA and EPA requirements and to reduce the risk of losses to your company.
Softcover, 192 pages, Nov '95, ISBN: 0-86587-505-7 **$59**

Safety Made Easy: A Checklist Approach to OSHA Compliance

This new book provides a simpler way of understanding your requirements under the complex maze of OSHA's Safety and Health regulations.
Softcover, 192 pages, June '95, ISBN: 0-86587-463-8 **$45**

Government Institutes • 4 Research Place • Rockville, MD 20850 • USA • (301) 921-2355